SpringerBriefs in Electrical and Computer Engineering

W0090525

For further volumes:
http://www.springer.com/series/10059

Jafar A. Alzubi · Omar A. Alzubi
Thomas M. Chen

Forward Error Correction Based On Algebraic-Geometric Theory

 Springer

Jafar A. Alzubi
Faculty of Engineering
Al-Balqa Applied University
Al-Salt
Jordan

Omar A. Alzubi
Prince Abdullah Ben Ghazi Faculty
  of Information and Technology
Al-Balqa Applied University
Al-Salt
Jordan

Thomas M. Chen
School of Mathematics, Computer Science
  and Engineering
City University London
Northampton Square
UK

ISSN 2191-8112   ISSN 2191-8120 (electronic)
ISBN 978-3-319-08292-9   ISBN 978-3-319-08293-6 (eBook)
DOI 10.1007/978-3-319-08293-6
Springer Cham Heidelberg New York Dordrecht London

Library of Congress Control Number: 2014942050

Printed on acid-free paper

Springer is part of Springer Science+Business Media (www.springer.com)

To Fatima and Mira

—Jafar A. Alzubi

To Zulfah, Mariam, Ahmad and Yousef

—Omar A. Alzubi

To Robin and Kayla

—Thomas M. Chen

Preface

Algebraic-geometric (AG) codes are a new paradigm in coding theory which promise performance improvements for point-to-point communications systems. AG codes offer several advantages over state-of-the-art Reed–Solomon (RS) codes. First, their construction is based on selecting points on a curve creating a non-binary code with long code length and effective decoding. The bit error rate (BER) performance of AG codes is impressive and attractive for wireless networks with severe fading conditions. Second, AG codes are more flexible than RS codes because they are easily extendable to high finite fields with minimal additional complexity. Third, the decoding approach gets all required information from the received data without the need for a decoding list. It is very attractive from the perspectives of both reliability and buffering capacity. Finally, construction of AG codes from curves offers an endless supply of AG codes with different properties and parameters applicable for different applications.

In this book, AG codes are designed, constructed and implemented from Hermitian curves. Simulations were carried out in Matlab to make comparisons of BER performance of AG codes and RS codes using different modulation schemes and various channel models such as additive white Gaussian noise (AWGN) and Rayleigh fast fading. Simulation results of BER performance for AG codes using quadrature amplitude modulation (16QAM and 64QAM) schemes are presented for the first time (to our knowledge) and shown to outperform RS codes at various code rates. Results for the AWGN channel are presented in this book; results for the Rayleigh fast fading channel are contained in the first author's Ph.D. dissertation.

To further improve the BER performance, algebraic-geometric block turbo codes (AG-BTCs) are proposed and implemented in this book. Their design, construction and implementation are investigated. Their performance is evaluated by simulations in Matlab, and results are presented for the first time in the literature. They show significant performance improvements but at the expense of high system complexity due to the use of Chase-Pyndiah's algorithm for AG codes.

In order to reduce system complexity while maintaining high BER performance, this book proposes algebraic-geometric irregular block turbo codes (AG-IBTCs). The design, construction and implementation of AG-IBTCs are presented along with new simulation results. Again appearing for the first time in the literature, results show that significant reduction in system complexity can be achieved while maintaining the high BER performance of AG-BTCs.

This book is intended to be useful to researchers and students in digital communications. The reader is assumed to have an appropriate background in mathematics and telecommunications. The presentation is intended to be self-contained with a substantial amount of background material included in the first half of the book. The second half concentrates on new research results. The advanced sections of the book may require a graduate level of education in communications.

This book is a result of the Ph.D. work carried out by the first author at the College of Engineering in Swansea University, Wales. The authors are grateful to Dr. Martin Johnston at Newcastle University for his invaluable assistance at the early stages of the research. Special thanks are given to Dr. Martin Crossley in the Mathematics Department at Swansea University for mathematical assistance throughout this work.

Salt, Jordan, April 2014   Jafar A. Alzubi
London, UK   Omar A. Alzubi
                                                                                              Thomas M. Chen

Contents

Acronyms

3G	Third Generation
ADSL	Asymmetric Digital Subscriber Line
AG	Algebraic Geometry
AG-BTC	Algebraic Geometric Block Turbo Code
AG-IBTC	Algebraic Geometric Irregular Block Turbo Code
AWGN	Additive White Gaussian Noise
BCH	Bose–Chaudhuri–Hocquenghem
BER	Bit Error Rate
BPSK	Binary Phase Shift Keying
BTC	Block Turbo Code
CTC	Convolutional Turbo Code
DVB	Digital Video Broadcasting
GF	Galois Field
HIHO	Hard Input Hard Output
HSPA	High Speed Packet Access
IBTC	Irregular Block Turbo Code
IDFT	Inverse Discrete Fourier Transform
ITC	Irregular Turbo Code
LDPC	Low Density Parity Check
LFSR	Linear Feedback Shift Register
LLR	Log-Likelihood Ratio
LR	Least Reliable
LTE	Long-Term Evolution
MAP	Maximum a Posteriori
ML	Maximum Likelihood
OFDMA	Orthogonal Frequency Division Multiple Access
PCCC	Parallel Concatenated Convolutional Code
PGZ	Peterson–Gorenstein–Zierler
QAM	Quadrature Amplitude Modulation
QPSK	Quadrature Phase-Shift Keying

RSC	Recursive Systematic Convolutional
SIHO	Soft Input Hard Output
SISO	Soft Input Soft Output
TPC	Turbo Product Code
VDSL	Very High Rate Digital Subscriber Line
WMAN	Wireless Metropolitan Area Network

Chapter 1
Introduction

In the past decade, the number of mobile devices has escalated driven mostly by demand for bandwidth-hungry smart phones. The need for efficient and reliable wireless communications has never been greater. The future Internet of Things (IoT) consisting of interconnected common objects capable of sensing and processing may generate orders of magnitudes more data. At the same time, the amount of radio spectrum is essentially limited, motivating a perpetual search for efficient coding schemes. Although major advances have been realized in coding, wireless mobile systems remain highly susceptible to impairments in the radio channel, and the control of transmission errors continues to be a major research problem and practical concern for communications system designers [1].

The basic principles of digital communication systems may be traced to Shannon's historic 1948 paper establishing the foundations of information theory [2]. This chapter was concerned with the transmission of symbols from an information source to a destination through a noisy channel. Following a probabilistic view of the information source, Shannon's source coding theorem established the concept of entropy as the lower limit on average bit rate for lossless source coding.

Shannon's noisy channel coding theorem described the maximum possible efficiency of error-correcting codes for a noisy channel. Channel capacity is the mutual information between the input and output of the channel maximized with respect to the input distribution. If the source information is transmitted at a rate less than the channel capacity, then there exist codes that allow the probability of error at the destination to be arbitrarily small. In other words, it is theoretically possible to transmit information with very low error at a rate up to the channel capacity. Conversely, if the transmission rate is more than the channel capacity, it is not possible to achieve an arbitrarily small error probability.

Since Shannon's contribution, the research community has worked diligently towards the goal of efficient encoding and decoding methods to control errors due to the noisy channel. Modern communication systems are typically designed with error control as an essential part. Continual advances in error control coding have led to more efficient and reliable digital communication systems.

J. A. Alzubi et al., *Forward Error Correction Based On Algebraic-Geometric Theory*,
SpringerBriefs in Electrical and Computer Engineering,
DOI: 10.1007/978-3-319-08293-6_1, © The Author(s) 2014

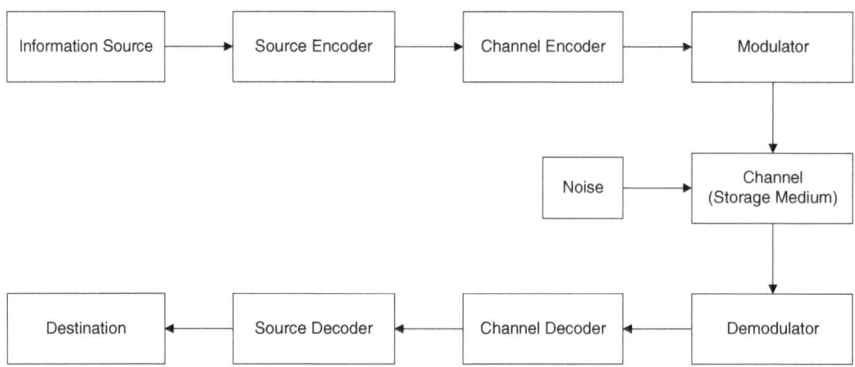

Fig. 1.1 A typical digital communication system

1.1 Digital Communications Systems

A classical view of a typical digital communication system is shown in the block diagram in Fig. 1.1 [1]. Generally, the information source could be analog or discrete. An analog source is usually assumed to be converted into a discrete source through analog-to-digital conversion consisting of sampling and quantization. A discrete source can transmit a sequence of symbols chosen from a known discrete alphabet. The source coder attempts to map the source symbols into bits as efficiently as possible, commonly by means of variable length coding. The process is sometimes called data compression. The idea of variable length coding is to assign shorter codewords to symbols that are more likely to be transmitted, and longer codewords to less likely symbols, thereby minimizing the average codeword length, e.g., by the well known Huffman code. The source coder produces a string of bits to the channel encoder.

The channel encoder and modulator depend on the characteristics of the channel. It is possible to simply use modulation without a channel encoder. Transmission is a physical process that is handled by the modulator. Without the channel encoder, the modulator converts bits from the source coder to baseband waveforms. If the channel is noiseless, the demodulator would convert the baseband waveforms back into bits for the source decoder to recover the transmitted symbols.

Unfortunately, there is no perfect (error-free) channel in actuality, and different types of media have different characteristics. Even optical fiber which is well known to be one of the best transmission media still has a very low bit error rate. The fiber acts as a waveguide for photons that is immune to external electromagnetic interference. The main causes of signal attenuation are light scattering and absorption within the fiber core. At the other extreme, radio channels are known to be one of the noisiest transmission media because they are vulnerable to several types of impairments such as reflections from objects (buildings, earth, atmospheric layers), diffraction (secondary waves bending around sharp obstructions), scattering, diffusion, attenuation, and multipaths (radio signals taking different paths to the receiver). In addition, the source and destination may be mobile and moving.

A critical part of communication system design is mathematical characterization of the channel. A common mathematical model is additive white Gaussian noise (AWGN) in which the impairment to communication is a linear addition of wideband or white noise with a constant spectral density (expressed as watts per hertz of bandwidth) and a Gaussian distribution of amplitude. The model does not account for fading, frequency selectivity, interference, nonlinearity or dispersion. However, it is popular due to its simplicity and tractability.

The AWGN channel is a good model for many satellite and deep space communication links. It is not a good model for most terrestrial links because of multipath, terrain blocking, interference, and so on. However, for terrestrial path modeling, AWGN is commonly used to simulate background noise of the channel under study.

1.1.1 Error Control Coding

In the presence of a noisy channel, the channel encoder becomes necessary for error control. Channel coding adds redundant bits after source coding to compensate for possible bit errors due to the imperfect channel. The channel encoder transforms the information sequence from the source encoder into a coded sequence of codewords. Codewords can be a binary or non-binary sequence. An enormous body of theory has been developed with many techniques for error control coding [3]. Common techniques include parity bits, cyclic redundancy checks (CRC), block codes (including Hamming, Reed-Solomon, Golay, BCH), and convolutional codes.

The channel decoder transforms the received sequence (of possibly corrupted codewords) into a binary or non-binary sequence called the estimated information sequence. The two main factors affecting decoding strategies are: the rules used in the channel encoding process and the noise characteristics of the channel (or storage medium).

A perfect channel encoding and decoding system will produce an estimated information sequence that is identical to the original information sequence, even though a number of decoding errors may introduced by the channel noise. The design and implementation of channel decoders is a major area of research since it plays a critical role in the performance of digital communication systems. Design of efficient channel decoders is an important topic in this book as well.

Design is governed by these considerations: the probability of decoding errors should be minimized; the transmission of information should be dense or fast as possible; the reproduced information at the channel decoder output should be reliable; and the implementation cost of the encoder and decoder should be reasonable [4].

1.1.2 Block and Convolutional Codes

Error control codes are divided structurally into two types: block codes and convolutional codes. The main difference between the two types is whether the encoder uses only the symbols in the current frame to produce its output as in block codes, or remembers a number of previous frames to produce its output as in the case of convolutional codes.

A new type of channel codecs was introduced in 1993 by Claude Berrou and Alain Glavieux called turbo codes (TCs) and block turbo codes (BTCs) which proved to be very powerful error correction techniques that outperformed all previously known coding schemes. They can be used in any communication system where a significant power saving is required or the operating signal-to-noise ratio (SNR) is very low. Deep space communications, mobile satellite/cellular communications, microwave links, and paging are some of the possible applications of this coding technique. The idea behind TCs can be thought of as a refinement of the concatenated encoding structure plus an iterative algorithm for decoding the associated code sequence [5].

A new family of non-binary block codes called algebraic-geometric (AG) codes were first introduced by V. D. Goppa in 1981. These codes are constructed from algebraic curves (e.g., Hermitian curves, elliptic curves, hyperelliptic curves) over finite fields. One property of the AG codes is that they have relatively long size [6].

One of the first and best known decoding algorithms for non-binary codes is the Berlekamp-Massey (BM) algorithm which proved to be very effective for short codes. However, because the decoding process involves two matrix inversions, the algorithm suffers from high complexity when dealing with long codes such as AG codes. To overcome this drawback, a new decoding algorithm essentially extending Berlekamp-Massey's algorithm was introduced by Sakata in 1988 [7]. Sakata replaced the matrix inversion processes by generating a set of polynomials whose coefficients formed recursive relationships among an array of finite field elements. Sakata's algorithm has been used in our design of the BTCs and the irregular BTCs.

1.2 Motivations

The motivation of this book is to investigate the construction, decoding, implementation, and BER performance evaluation of AG codes. A well known construction method of AG codes presented by Justesen et al. is used in this book owing to its simplicity and versatility for different channel models. The constructed AG codes have shown significant improvements in BER performance in comparison to RS codes. This motivates us to use AG codes as code components of BTCs in the pursuit of further performance improvements.

One important characteristic of AG codes is that they produce hard output. This does not fit well with the concept of BTCs where a soft output is usually required. This motivates us to consider Chase-Pyndiah's approach for extracting soft output from hard decision output and then converting the one-pass system to an iterative system in order to improve the BER performance further.

Reed-Solomon block turbo codes (RS-BTCs) have been used as a reference to measure the gain in BER performance of AG-BTCs. In addition to the BER performance, the complexity of the decoding process is an important trade-off with the performance. However, using AG codes along with Chase-Pyndiah's algorithm may lead to an increase in the decoding complexity for better performance.

In order to reduce the decoding complexity of the resultant system and consider practical implementation, we design algebraic-geometric block turbo codes (AG-BTCs) by constructing suitable algebraic-geometric irregular block turbo codes (AG-IBTCs). The construction, decoding and implementation of the new IBTC are investigated here. The performance of the new constructed AG-IBTCs is compared with the performance of the equivalent AG-BTC over different channel models and several modulation schemes.

1.3 Aims and Objectives

This book aims to design, construct and implement a reliable communications system with relatively low complexity compared to state-of-the-art systems. The design, construction and implementation of AG codes for use as code components in BTCs and IBTCs are investigated. The BER performance of various AG codes are compared with the equivalent Reed-Solomon (RS) variations of BTC by means of computer simulations. Comparison results are presented for several code rates and modulation schemes over various practical channel models.

The objectives of this book can be summarized as:

- Design and construct long AG codes and compare their BER performance with equivalent RS codes;
- Construct a new BTC by employing Chase-Pyndiah's algorithm for extracting soft outputs from hard decision outputs using the AG codes as code components;
- Evaluate the BER performance of the new AG-BTCs in comparison with RS-BTCs by means of computer simulations;
- Design and construct IBTCs using AG codes as code components in order to reduce the decoding complexity of AG-BTCs and enhance the BER performance as possible;
- Evaluate the BER performance of the new AG-IBTCs in comparison with equivalent AG-BTCs through computer simulations;
- Evaluate the above constructed codes over AWGN channels using several modulation schemes through computer simulations.

1.4 Original Contributions

In this book, AG codes are constructed using the simplified method of Justesen et al. from Hermitian curves over the finite field $GF(2^4)$ with varying code rates. The extension of BM decoding algorithm presented by Sakata is used with a majority voting (MV) technique to decode the produced codes. The performance of the constructed codes is evaluated in terms of BER over AWGN channel with binary phase-shift keying (BPSK) modulation scheme which matches the well known results in the literature. Moreover, the first simulation results showing the performance of AG codes over AWGN channel using quadrature phase-shift keying (QPSK), 16 QAM and 64 QAM modulation schemes are presented.

In addition, an AG iterative decoding technique is developed for non-binary BTCs constructed from AG codes as code components. Iterative decoding is applied to AG codes in order to enhance performance. This is done with the use of Chase-Pyndiah's decoding algorithm for extracting soft output from a hard decision output (AG decoder based on Sakata's algorithm). Simulation results show that AG-BTCs outperform the RS-BTCs in AWGN channels over the above mentioned modulation schemes.

In order to reduce system complexity, AG-IBTCs are proposed, designed, and constructed. Measurements of the BER performance of the designed AG-IBTCs show that they perform no worse than the regular AG-BTCs and frequently better especially at higher order modulation schemes. Moreover, the AG-IBTCs system's complexity is always reduced significantly compared to the complexity of AG-BTCs.

1.5 Book Layout

This book is organized into five chapters as follows:

- Chapter 1:
  This chapter motivates the book, provides an overview, lists objectives and aims, and summarizes the key contributions of the work.
- Chapter 2:
  This chapter presents the theoretical background covering AG code creation, encoding and decoding as well as fundamentals of TCs and BTCs.
- Chapter 3:
  This chapter reviews the literature on AG codes construction and decoding methods and the decoding of regular and irregular BTCs.
- Chapter 4:
  This chapter extends the AG codes design into BTCs design by using the AG codes as code components of BTCs and shows AG codes construction using Justesen's simplified method. Also the AG iterative decoding technique using Chase-Pyndiah's algorithm is presented in this chapter.

- Chapter 5:
  This chapter introduces the developed IBTC with AG codes as code components.
- Chapter 6:
  This chapter finally discusses all the results achieved and draws conclusions with a discussion of possible future work.

References

1. Lin S, Costello DJ (2004) Error control coding, 2nd edn. Prentice-Hall, Upper Saddle River
2. Shannon CE (1948) A mathematical theory of communication. Bell Syst Tech J 27(3):379–423
3. Peter-Sweeney P (2004) Error control coding: from theory to practice. Wiley, New York
4. Sklar B (1988) Digital communications: fundamentals and applications. Prentice-Hall, Upper Saddle River
5. Vucetic B, Yuan J (2000) Turbo codes: principles and applications. Kluwer Academic, Boston. http://www.loc.gov/catdir/enhancements/fy0820/00033104-t.html
6. Biglieri E (2005) Coding for wireless channels. Information technology-transmission, processing, and storage. Springer, Berlin. http://books.google.co.uk/books?id=seoUuxiqG-oC
7. Sakata S (1988) Finding a minimal set of linear recurring relations capable of generating a given finite two-dimensional array. J Symbolic Comput 5(3): 321–337. doi:10.1016/S0747-7171(88)80033-6. http://www.sciencedirect.com/science/article/pii/S0747717188800336

Chapter 2
Theoretical Background

In this chapter, the theoretical background is presented covering design and construction of AG codes for the encoder and decoder along with important parameters. We also present a block diagram of the modified Sakata's algorithm for the first time. It shows how the construction of AG codes using Hermitian codes is performed using a hard-input hard-output (HIHO) decoding algorithm. Fundamentals of TCs encoder, decoder and interleaver design are shown. Examples of the construction of BTCs are also presented.

2.1 Algebraic Geometric Codes

For a long time researchers attempted to realize a very long non-binary block code with high code rate and large Hamming distance, however fulfilling these properties by classical codes was not possible. In 1981, V. D. Goppa showed a way to construct these codes which are now called Goppa codes or AG codes [1]. Goppa explained the construction from affine points of an irreducible projective curve and a set of rational functions defined on that curve. The famous Reed-Solomon (RS) code represents the best and for most the simplest example that demonstrates the construction of AG codes though it is constructed from the affine points of a projective line not a projective curve which is the case of Goppa codes.

The number of affine points determines the length of an AG code, so the cardinality of the chosen field restricts the length of RS codes which result in relatively short code lengths. Replacing the projective line with a projective curve yields more affine points which means longer code lengths while keeping the same size of the finite field [2, 3]. The longest possible codes can be obtained by choosing curves that have the maximum number of affine points which are called maximal curves, so the objective is always to find those curves whenever possible.

A possible reason that AG codes have not been studied and investigated very well is that they require a good knowledge of the theory of algebraic geometry, a difficult

J. A. Alzubi et al., *Forward Error Correction Based On Algebraic-Geometric Theory*,
SpringerBriefs in Electrical and Computer Engineering,
DOI: 10.1007/978-3-319-08293-6_2, © The Author(s) 2014

and complicated branch of mathematics. To overcome the previously stated problem, a simplified construction method was introduced in 1989 by Justesen et al. [4]. His method requires a basic understanding of algebraic geometry to produce AG codes. Although a limited number of AG codes—which is considered as a drawback— can be constructed using this method compared with using a more complicated AG approach, however this limited number of codes is still acceptable.

2.1.1 Construction of AG Code Parameters

According to Carrasco [5], an AG code can be constructed using Justesen's simplified method by choosing an irreducible affine smooth curve over a finite field. Classes of good curves that could be used to produce good AG codes are the Hermitian curves, elliptic curves, hyperelliptic curves, and so on, as they all have one point at infinity.

However, Hermitian curves with degree $m = r + 1$ where $r = \sqrt{q}$ are well known from the previous classes of curves and most popular for constructing AG codes defined over a finite field F_q [4]:

$$C(x, y) = x^{r+1} + y^r + y \tag{2.1}$$

To define the message length (k) and the designed minimum Hamming distance (d^*), all affine points (the points causing the curve to vanish) as well as the point at infinity on the chosen curve must be found. The number of the affine points which satisfy $C(x, y) = 0$ is $n = r^3$. Hasse-Weil bound gives an upper bound for the number of affine points n [4]:

$$n \leq 2\gamma\sqrt{q} + 1 + q \tag{2.2}$$

where γ is the genus of the curve.

It is worth giving a complete explanation of the curve genus as it is difficult to find a detailed simplified definition and method of genus computation. The genus is the maximum number of cuttings along non-intersecting simple curves [6]. The process of computing it is perhaps more interesting. Assume there exists a plane curve called C which is defined by $f(x, y) = 0$ where $f(x, y)$ is a two-dimensional polynomial composed of two variables. The degree of this polynomial is m which is the largest sum of the exponents of x and y in each term of the curve equation. Then the genus of C is:

$$\gamma = \frac{(m - 1)(m - 2)}{2} \tag{2.3}$$

if and only if C is non-singular curve.

A nonsingular curve, also called smooth curve, is the one which has no singular points. A singular point is defined as the point where something unusual happens on the curve like a sharp corner ($y^2 = x^3$) or a crossing of two branches ($y^2 = x^3 + x^2$).

Otherwise, when the curve has a finite number of singular points, it is called a singular curve [7].

As Hermitian curves saturate the Hasse-Weil bound, they called maximal curves making them suitable to generate long AG codes. Justesen's construction method suggests a non-negative integer j which is bounded by [8]:

$$m - 2 \leq j \leq \left| \frac{n-1}{m} \right| \tag{2.4}$$

Goppa or AG codes are of two types: functional Goppa codes (C_L) and residue Goppa codes (C_Ω). The latter is the dual of the former. In both types, the block length is equal to the number of affine points on the curve (n) [5]. To compute the length of the message for an AG code, a set of rational functions with a pole of order equal or less than the degree of the divisor (a) at the point at infinity (Q) must be found first [6], where the degree of the divisor is limited to be greater than $2\gamma - 2$ and less than n ($2\gamma < a < n$). In Justesen's simplified construction method $a = mj$. This set of rational functions is also called the linear space of aQ which is denoted by $L(aQ)$.

The number of elements in the previous set is equal to the message length k. It is called the dimension of aQ and denoted by $l(aQ)$ [8]. The Riemann-Roch theorem is used to calculate $l(aQ)$ [9, 10] which defines the message length k in functional Goppa codes $C_L(D, aQ)$ as:

$$k = l(aQ) = deg(aQ) - \gamma + 1 = a - \gamma + 1 \tag{2.5}$$

while the message length k in residue Goppa codes is defined by:

$$k = n - l(aQ) = n - a + \gamma - 1 \tag{2.6}$$

For both types of AG codes, a lower bound of the Hamming distance of AG codes is calculated and called the designed minimum Hamming distance d^* as the Hamming distance (d) cannot always be calculated accurately. Meeting the singleton bound when calculating minimum Hamming distance is required as the value will then be optimal [11]:

$$d = n - k + 1 \tag{2.7}$$

However, the main disadvantage that must be mentioned regarding the use of AG codes is that the designed minimum Hamming distance is affected inversely by the genus of the curve. This means that the larger the genus, the smaller the designed minimum Hamming distance, and vice versa. In contrast, the case of RS codes are constructed over an affine line of degree one and genus equal to zero [5].

So the actual designed minimum Hamming distance is [8, 10]:

$$d^* = n - k - \gamma + 1 \tag{2.8}$$

To compute the designed minimum Hamming distance for functional Goppa codes, we substitute the value of message length for those codes into (2.8) and get that:

$$d^* = n - a \qquad (2.9)$$

Also for residue Goppa codes, the designed minimum Hamming distance can be found by substituting the message length k into (2.8) and get:

$$d^* = a - 2\gamma + 2 \qquad (2.10)$$

As Justesen's simplified code have the same parameters as residue Goppa codes since $a = mj$, then the code parameters are [8]:

$$K = n - mj + \gamma - 1 \qquad (2.11)$$

$$d^* = mj - 2\gamma + 2 \qquad (2.12)$$

and the codeword length n is equal to the number of affine points on the curve as mentioned earlier.

2.1.2 Designing AG Encoder

To generate a generator matrix for an AG code, all the points that satisfy the chosen curve must be found which means all the points that make $C(x, y) = 0$ excluding the point at infinity. For Harmitian curves, as previously said, the number of these points is equal to $n = r^3$ where $r = \sqrt{q}$, and q is the finite field size [8].

A k two variables monomial basis is defined as: $F = x^a y^b$ where $0 \le a < m$ and $b \ge 0$ and ordered using total graduated degree ($<_T$). This method of ordering follows the pattern: first-degree pair $(a, b) = (0, 0)$; next-degree pair (a', b') is [12]:

$$(a', b') = \begin{cases} (a - 1, b + 1) & \text{if } a > 0 \\ (b + 1, 0) & \text{if } a = 0 \end{cases} \qquad (2.13)$$

So, degree pairs ordering is: $(0, 0) <_T (1, 0) <_T (0, 1) <_T (2, 0) <_T (1, 1) <_T (0, 2) <_T (3, 0) <_T (2, 1) <_T (1, 2) <_T (0, 3) <_T (4, 0) <_T (3, 1) <_T (2, 2)\ldots$ This gives monomial basis (ϕ_i):

$$\left\{ 1, x, y, x^2, xy, y^2, x^3, x^2y, xy^2, y^3, x^4, x^3y, x^2y^2, xy^3, y^4, x^5, \ldots \right\} \qquad (2.14)$$

It is worth explaining another ordering technique which is called partial ordering as it will help to show the concrete difference between the two ordering techniques and will be helpful in understanding steps of the decoding procedure later on. Assume there are two pairs of integers $a = (a_1, a_2)$ and $b = (b_1, b_2)$ then [12]:

$$a < b \quad \text{if} \quad a_1 \leq b_1 \wedge a_2 \leq b_2 \wedge a \neq b \tag{2.15}$$

To obtain the final non-systematic generator matrix of the code, each of the monomial basis ϕ_i, $i = 1, 2, ..., k$ in $L(aQ)$ is evaluated at each affine point as the following:

$$G = \begin{bmatrix} \phi_1(p_1) & \phi_1(p_2) & \cdots & \phi_1(p_{n-1}) & \phi_1(p_n) \\ \phi_2(p_1) & \phi_2(p_2) & \cdots & \phi_2(p_{n-1}) & \phi_2(p_n) \\ \phi_3(p_1) & \phi_3(p_2) & \cdots & \phi_3(p_{n-1}) & \phi_3(p_n) \\ \vdots & \vdots & \ddots & \vdots & \vdots \\ \phi_{k-1}(p_1) & \phi_{k-1}(p_2) & \cdots & \phi_{k-1}(p_{n-1}) & \phi_{k-1}(p_n) \\ \phi_k(p_1) & \phi_k(p_2) & \cdots & \phi_k(p_{n-1}) & \phi_k(p_n) \end{bmatrix} \tag{2.16}$$

Extracting the original message from the decoded codeword is a difficult and complex process when working with a non-systematic generator matrix. Multi-stage shift register technique is used in cyclic codes like RS codes to produce systematic generator matrix from a non-systematic one [5]. However, the technique does not work for AG codes since they are not cyclic, so another technique called Gauss-Jordan elimination could be used to convert the non-systematic generator matrix to a systematic one, keeping in mind that any interchange in columns while applying Gauss-Jordan elimination must be followed by same pattern on points [8, 13].

2.1.3 Designing AG Decoder

The traditional decoding technique for RS codes consists of two stages: the purpose of the first stage is to find the error locations while the second stage attempts to compute the error magnitudes for each of the found locations. AG codes follow the previously described technique [14].

In 1969, the BM algorithm [15] was introduced as a way to produce a shortest linear feedback shift register (LFSR) which yields a finite sequence of digits. By using the BM algorithm in 1988, Sakata was able to develop his algorithm which generates a set of minimal polynomials whose coefficients form a recursive relationship within a two-dimensional array of finite field elements [12].

Justesen et al. [16] were able to improve Sakata's algorithm in 1992. The aim of this improvement was to decrease the decoding complexity of AG codes by generating a set of error-locating polynomials (F) from a two-dimensional matrix containing syndrome values for AG codes. The decoding process starts with computing the elements in the two-dimensional syndromes array. Let us refer to the element location in the two dimensional array by $(S_{a,b})$ where a is the row number and b is the column number $(a, b < q - 1)$. The syndrome value is defined by [16]:

$$S_{a,b} = \sum_{i=1}^{n} r_i x_i^a y_i^b = \sum_{i=1}^{n} (c_i + e_i) x_i^a y_i^b = \sum_{i=1}^{n} e_i x_i^a y_i^b \tag{2.17}$$

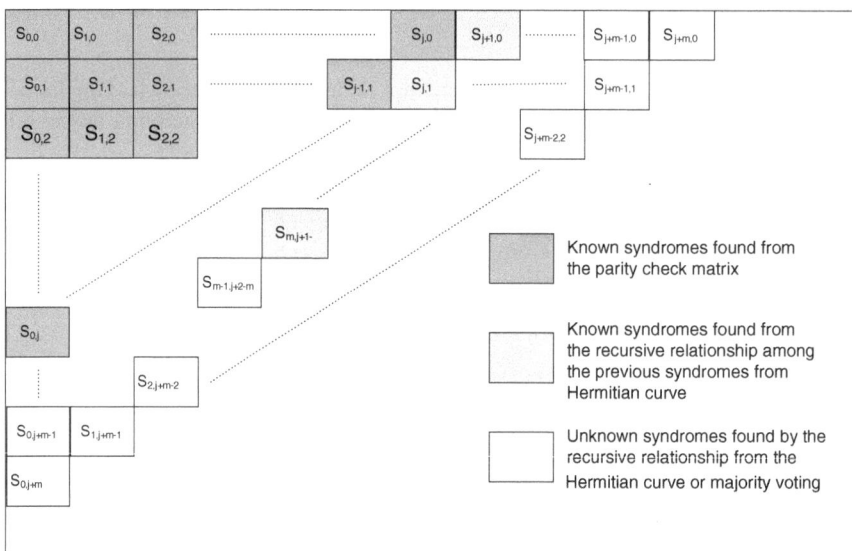

Fig. 2.1 General two-dimensional syndrome array

Let r_i be a received element within the received codeword r, c_i be a coded symbol, e_i the corresponding error magnitude in the i-th position, and (x_i, y_i) the i-th affine point, for $i \in I$, $I \subseteq \{1, 2, 3, ..., n\}$. The general two-dimensional syndrome array for the AG code constructed from the Hermitian curve defined by (2.1) is shown in Fig. 2.1 [5].

Sakata's algorithm makes use of this two-dimensional array by creating a set of error-locating polynomials (F) of the form [12]:

$$f^{(i)}(x, y) = \sum f_{k,l}^{(i)} x^k y^l \qquad (2.18)$$

where the i-th polynomial in F is denoted by i, and the coefficients of the terms $x^k y^l$ in $f^{(i)}(x, y)$ is represented by $f^{(i)}(k, l)$. By reading every syndrome value in the two-dimensional array using the total graduate degree order ($<_T$), all polynomials in F are updated in order to generate recursive relationships between known syndromes, up to the current syndrome by changing all the coefficients of every polynomial $f^{(i)}(x, y)$ [17]. However the generated recursive relationship needs to fulfill the following equation [5]:

$$\sum f_{k,l}^{(i)} S_{a-t_1^{(i)}+k, b-t_2^{(i)}+l} = 0 \qquad (2.19)$$

where $f^{(i)}(x, y)$ is a polynomial in the set F and has x, y as variables of the leading term with $t_1^{(i)}$ and $t_2^{(i)}$ representing their powers, respectively. Using the total graduated degree ordering ($<_T$) described previously, the syndromes in the two-dimensional array are read as following: $S_{0,0}$, $S_{1,0}$, $S_{0,1}$, $S_{2,0}$, $S_{1,1}$, $S_{0,2}$, $S_{3,0}$, $S_{2,1}$, and so on until the last syndrome in the array which is $S_{15,15}$.

The nonnegative integer j defined by (2.4) plays an important role in articulating the syndromes. However, the syndromes are categorized into two types: known and unknown, where the known ones are from $S_{0,0}$ to $S_{m,j+1-m}$. To compute the syndromes $S_{0,0}$ to $S_{0,j}$ Eq. (2.17) is used. By substituting the curve equation in the equation representing the error-locating polynomial in (2.18), the following recursive relationship is formed to calculate more known syndromes $S_{j+1,0}$ to $S_{m,j+1-m}$ [12]:

$$\sum_{k,l} C_{k,l} S_{a-t_1^{(i)}+k, b-t_2^{(i)}+l} = 0 \qquad (2.20)$$

$$C_{0,1} S_{a-m+0, b-0+1} + C_{0,m-1} S_{a-m+0, b-0+m-1} + C_{m,0} S_{a-m+m, b-0+0} = 0 \quad (2.21)$$

where the coefficient of $C(x, y)$ is $C_{k,l}$, and the powers of x and y for each term in $C(x, y)$ are k and l, respectively. This relationship could be simplified into the following equation as all coefficients of $C(x, y)$ are equal to one [5]:

$$S_{a,b} = S_{a-m, b+1} + S_{a-m, b+m-1} \qquad (2.22)$$

Next is the time to update the set F by testing all polynomials $(f^{(i)}(x, y) \in F)$ to see whether they satisfy (2.19). If they do, then none needs to be changed. Otherwise, if any of these polynomials do not satisfy (2.19) then this polynomial will be used in the updating process of the set F because it has a discrepancy d_f. This means that the polynomial at this stage is not ideal and must be changed so that it satisfies (2.19) after updating. The goal is to have a set of error-locating polynomials in F, and a polynomial is said to be error-locating if and only if it satisfies (2.19) [5, 8, 12].

The polynomials that do not satisfy (2.19) by having a nonzero discrepancy will be placed in a new set called auxiliary set (G). Also the point at which they were placed is stored (a_g, b_g). At this stage of the decoder, a new set $span(G)$ is generated by the union of all sets less than or equal to each $span(g^{(i)}(x, y))$ in G as in following Equation [5]:

$$span(G) = \sum_{i=1}^{\varphi} \{(k, l) \mid (k, l) \leq span(g^{(i)}(x, y))\} \qquad (2.23)$$

where (k, l) are a pair of positive integers and φ is the number of polynomials in the set G. Span means that at the point (a, b) there is no polynomial with a leading term $x^{a_g - u_1^{(i)}} y^{b_g - u_2^{(i)}}$ that can satisfy (2.19). It is defined as [12]:

$$span(g^{(i)}(x, y)) = (a_g - u_1^{(i)}, b_g - u_2^{(i)}) \qquad (2.24)$$

where $g^{(i)}(x, y)$ is a polynomial in the set G and has x, y as variables of the leading terms with $u_1^{(i)}$ and $u_2^{(i)}$ representing their degrees, respectively.

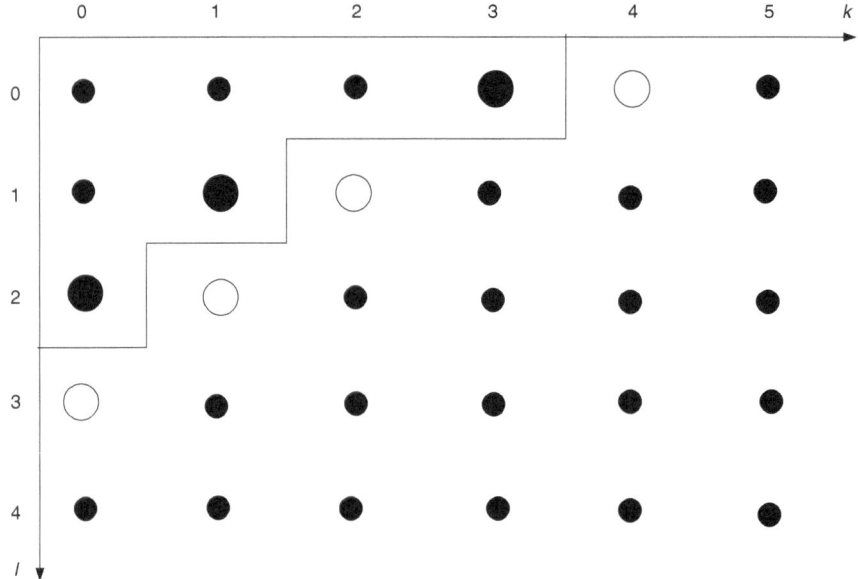

Fig. 2.2 Graphical representation of $span(G)$

The maximal point within $span(G)$ is defined as an interior corner while the minimal point outside $span(G)$ is called as exterior corner [5]. Both interior and exterior corners are defined with respect to partial ordering which is denoted by $(<)$ as mentioned earlier. However, the values of the exterior corners are the degrees of the polynomials in the set F and their number is the number of those polynomials.

Drawing $span(G)$ makes it much easier to find the interior and exterior corners. An example shows how drawing helps in finding out these corners. Assume $span(G) = \{(0, 0), (1, 0), (0, 1), (2, 0), (1, 1), (0, 2), (3, 0)\}$. From Fig. 2.2, the exterior corners are $(4, 0)$, $(2, 1)$, $(1, 2)$, and $(0, 3)$ as no other points outside $span(G)$ are less than them. In the same manner, it is shown that the interior corners are $(3, 0)$, $(1, 1)$, and $(0, 2)$ since these points are the greatest ones within $span(G)$. Exterior corners are marked with large white circles in the figure and the interior corners are marked with large black circles [5].

2.1.3.1 Updating the Sets F and G

The polynomials in the set F with a nonzero discrepancy $(d_f \neq 0)$ are stored in a new set called F_N. The union of this set with the set G results a new set called G' which is the updated version of the set G $(G' = G \cup F_N)$ [8]. Equation (2.23) is used to calculate the span of each polynomial in G', then Eq. (2.22) is used to find $span(G')$. The interior corners are found then using $span(G')$ so that any polynomial in the set G' with span not equal to any of the interior corner will be removed. In case two or more polynomials in the set G' have span equal to any interior corner then any of those polynomials will be kept. The point in the two-dimensional syndrome array (a_g, b_g) where the discrepancies of remaining G' polynomials were nonzero and the discrepancy of the set G' polynomials d_g are stored [12]. G' at this stage is the final update of the set G which will be used for the next point in the two-dimensional syndrome array.

The exterior corners are found using $span(G')$ to update the set F. As mentioned earlier, the number of the exterior corners identifies the number of the polynomials in the updated set F'. Also their values are the powers of the leading terms of these polynomials [5]. The polynomials in the set F are updated using one of three cases for each of the exterior corners $(\varepsilon_1, \varepsilon_2)$, however these cases must be applied in the following order [8, 18]:

Case I If the difference set (F/F_N) has a polynomial $f^{(i)}(x, y)$ with $(t_1^{(i)}, t_2^{(i)}) = (\varepsilon_1, \varepsilon_2)$, then the new minimal polynomial $h^{(i)}(x, y) \in F'$ will be the same:

$$h^{(i)}(x, y) = f^{(i)}(x, y) \tag{2.25}$$

Case II If there is a polynomial $f^{(i)}(x, y) \in F_N$ with $(t_1^{(i)}, t_2^{(i)}) \leq (\varepsilon_1, \varepsilon_2)$ and $\varepsilon_1 > a$ or $\varepsilon_2 > b$, then the new minimal polynomial $h^{(i)}(x, y) \in F'$ is generated using:

$$h^{(i)}(x, y) = x^{\varepsilon_1 - t_1^{(i)}} y^{\varepsilon_2 - t_2^{(i)}} - f^{(i)}(x, y) \tag{2.26}$$

Case III If there is a polynomial $g^{(i)}(x, y) \in G$ having span greater than or equal to $(a - \varepsilon_1, b - \varepsilon_2)$ and a polynomial $f^{(i)}(x, y) \in F_N$ with $(t_1^{(i)}, t_2^{(i)}) \leq (\varepsilon_1, \varepsilon_2)$, then the new minimal polynomial $h^{(i)}(x, y) \in F'$ is generated using:

$$h^{(i)}(x, y) = x^{\varepsilon_1 - t_1^{(i)}} y^{\varepsilon_2 - t_2^{(i)}} f^{(i)}(x, y) - \frac{d_f}{d_g} x^{p_1} y^{p_2} g^{(i)}(x, y) \tag{2.27}$$

where $(p_1, p_2) = span(g^{(i)}(x, y)) - (a - \varepsilon_1, b - \varepsilon_2)$. Whenever, an update occurs to the set F, a new set denoted by Δ is developed. It will be used for the MV technique as part of the decoding procedure and also for termination of the decoding algorithm when required. It is defined as [8]:

$$\sum_{k=1}^{\lambda - 1} \Delta_k \tag{2.28}$$

where λ represents the number of all polynomials in the set F. Further, Δ_k is defined by:

$$\Delta_k = \left\{ (k, l) \mid (k, l) \leq \left(t_1^{(k)} - 1, t_2^{(k+1)} - 1 \right) \right\} \tag{2.29}$$

where (k, l) are a pair of positive integers.

A major modification was introduced to the decoding algorithm which was concerned with adding a termination criteria for the algorithm when $|\Delta|$ becomes greater than the number of errors that the decoder can handle [5, 8]. In such case, the MV scheme will choose a false value for the unknown syndrome which will affect the accuracy of finding the remaining unknown syndromes resulting in inaccurate decoding.

After completing the two-dimensional syndrome array, all polynomials in the set F are said to be error-correcting polynomials which means when substituting the curve points into any of those polynomials, the error locations are the points that make the polynomial vanish [18].

A modified version of Sakata's algorithm is illustrated in the flow chart shown in Fig. 2.3. The best of our knowledge, this flow chart is the first published illustration in the literature.

2.1.3.2 Majority Voting

Sakata's algorithm uses the technique of substituting the curve Eq. (2.1) into (2.19) to come up with a recursive relationship among the previous syndromes to find the unknown syndromes of the type $S_{a,b}$ where $a \leq m$. This can be true if and only if all previous syndromes are known. If any of those previous syndromes are unknown, then the MV technique is used to compute the unknown syndromes of the type $S_{a,b}$ where $a < m$ [5, 12].

The following example will help clarify the idea. For an algebraic geometric code constructed from a Hermitian curve of the form given in (2.1) with degree $m = 5$, Eq. (2.22) can be used to compute the syndrome $S_{8,1}$:

$$S_{8,1} = S_{8-5,1+1} + S_{8-5,5-1} \tag{2.30}$$
$$= S_{3,2} + S_{3,4} \tag{2.31}$$

This only holds if both $S_{3,2}$ and $S_{3,4}$ are known. Otherwise, MV technique is used to find the unknown ones.

In 1993 Feng and Rao [19] introduced the MV scheme which Sakata et al. used later in 1995 [17] to design a hard-decision decoding technique for AG codes. For an unknown syndrome of the type $S_{a,b}$ where $a < m$, any minimal polynomial $f^{(i)}(x, y) \in F$ will be used to find a candidate syndrome value. It turns out that there are four possible scenarios to be encountered depending on some conditions which will be explained in detail below[17].

Scenario one: The candidate syndrome value is v_i if $a = t_1^{(i)}$ and $b = t_2^{(i)}$ can be calculated by using the following equation which is derived from (2.19):

$$\sum_{(k,l) \leq T\left(t_1^{(i)}, t_2^{(i)}\right)} f_{k,l}^{(i)} S_{k+a-t_1^{(i)}, l+b-t_2^{(i)}} = -v_i \tag{2.32}$$

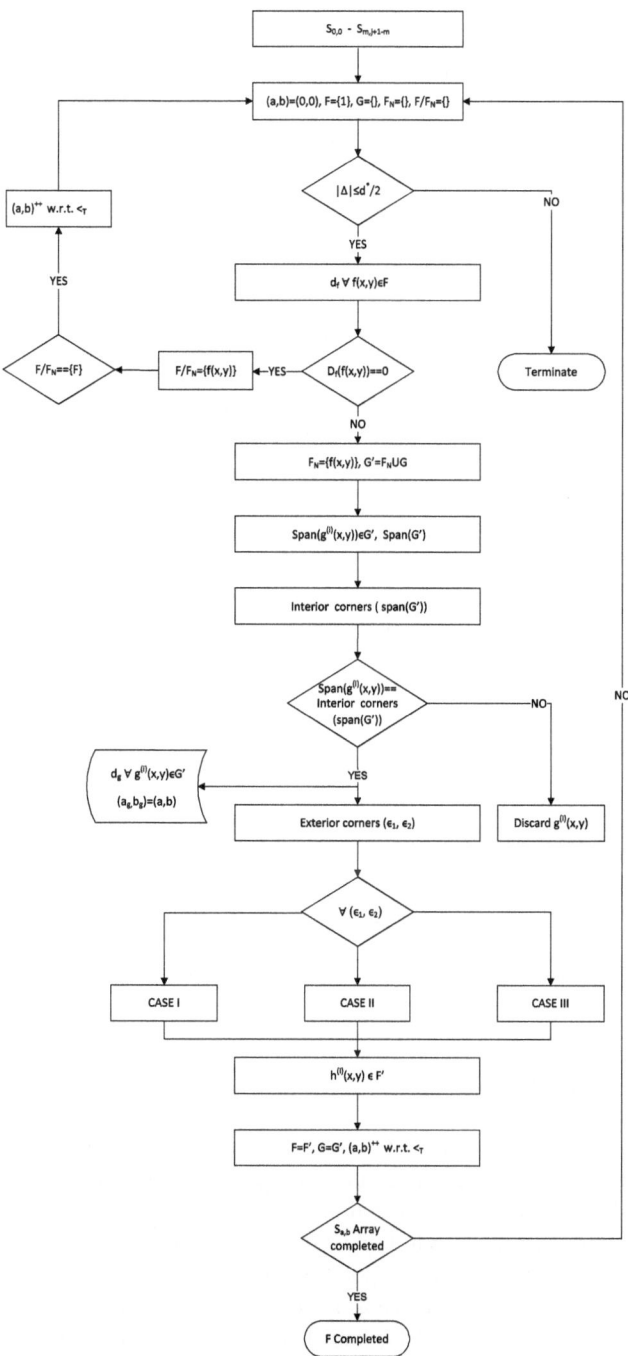

Fig. 2.3 Flow chart of modified version of Sakata's algrithm

Scenario two: The candidate syndrome value is w_i if $a + m = t_1^{(i)}$ and $b - t_2^{(i)} = m - 1$ can be calculated by using the following equation:

$$\sum_{(k,l) <_T \left(t_1^{(i)}, t_2^{(i)}\right)} f_{k,l}^{(i)} S_{k+a+m-t_1^{(i)}, l+b-m+1-t_2^{(i)}} - S_{a,b-m+2} = -w_i \qquad (2.33)$$

Scenario three: If both scenarios one and two are fulfilled which means (2.32) and (2.33) are satisfied, then there will be two candidate values v_i and w_i for the syndrome from this minimal polynomial.

Scenario four: If none of the above three scenarios are fulfilled, then the chosen minimal polynomial $f^{(i)}(x, y) \in F$ is not capable of finding a candidate syndrome value, so a different minimal polynomial $f^{(i)}(x, y) \in F$ will be considered.

The next step in calculating the MV is to generate two new sets:

$$K_1 = \{(k, l) \mid 0 \le k \le a \wedge 0 \le l \le b\}$$
$$K_2 = \{(k, l) \mid 0 \le k < m \wedge 0 \le l \le b - m + 1\} \qquad (2.34)$$

where (k, l) are a pair of positive integers. A set $K = K_1 \cup K_2$ is computed also. The MV decision is made based on the number of elements in the set K_j which is found for each candidate syndrome value $\delta_1, \delta_2, \delta_3, \ldots$ as [17]:

$$K_j = \left(\bigcup_{v_i = \delta_j} A_i \cup \bigcup_{w_i = \delta_j} B_i \right) \Big/ \Delta \qquad (2.35)$$

where A_i and B_i are defined by:

$$A_i = \left\{ (k, l) \in K \mid k + t_1^{(i)} \le a \wedge l + t_2^{(i)} \le b \right\}$$
$$B_i = \left\{ (k, l) \in K \mid k + t_1^{(i)} \le a + m \wedge l + t_2^{(i)} \le b - m + 1 \right\} \qquad (2.36)$$

2.1.3.3 Error Magnitudes

To find the error magnitude for AG codes (generated from Hermitain curves), the points on the curve are categorized into four types. The magnitude of the error will be calculated based on the error location on the curve which means it does matter where the error occurs in order to find its magnitude [8].

The following four categories of points on the curve will be useful in the method described below. The method depends on calculations of a one-dimensional inverse discrete Fourier transform (IDFT), knowledge of the unknown syndromes up to syndrome $S_{q-1,q-1}$, and the curve properties [17, 20].

Category one: For error occurring at the point where both coordinates x and y are zeros, i.e., $P_{(0,0)}$. The error value is found by subtracting the error values of errors which occurred at all other points types $P_{(x,0)}$, $P_{(0,y)}$, and $P_{(x,y)}$ as following:

$$\sum_{p_i \in P_{(x,y)}} e_i = S_{q-1,q-1}$$

$$\sum_{p_i \in P_{(0,y)}} e_i = S_{0,q-1} - S_{q-1,q-1} \tag{2.37}$$

$$\sum_{p_i \in P_{(x,0)}} e_i = S_{q-1,0} - S_{q-1,q-1}$$

This leads to:

$$e_i = \sum_i e_i - \sum_{p_i \in P_{(x,y)}} e_i - \sum_{p_i \in P_{(0,y)}} e_i - \sum_{p_i \in P_{(x,0)}} e_i$$

$$= S_{0,0} - S_{q-1,q-1} - (S_{0,q-1} - S_{q-1,q-1}) - (S_{q-1,0} - S_{q-1,q-1}) \tag{2.38}$$

However, for the codes constructed from the curves over a finite field of characteristic two, Eq. (2.38) can be simplified to:

$$e_i = S_{0,0} + S_{q-1,0} + S_{q-1,q-1} + S_{0,q-1} \tag{2.39}$$

Category two: For all errors occurring at the points of zero x-coordinate and nonzero y-coordinate, i.e., $P_{(0,y)}$, the following mapping is defined:

$$m \rightarrow \begin{cases} \alpha^m & \text{for } 0 \leq m \leq q - 2 \\ 0 & \text{for } m = q - 1 \end{cases} \tag{2.40}$$

and the one-dimensional IDFT equation is:

$$E_n = \sum_{i=0}^{q-2} S_{0,q-1-i} \alpha^{ni} \tag{2.41}$$

where α is the primitive element of the finite field and E_n is the summation of all error values occurred at the points of nonzero y-coordinate α^n. Luckily, Hermitian curves (the focus here) have a property that whenever there is a point on the curve of zero x-coordinate and nonzero y-coordinate, there will be no points on the curve with the same y-coordinate value with nonzero x-coordinate (α^m, α^n). Which means that E_n is in fact the error magnitude of the error at the point $P_{(0,y)} = (0, \alpha^n)$.

Category three: For all errors occurring at the points of nonzero x-coordinate and zero y-coordinate, i.e., $P_{(x,0)}$, the same mapping (2.40) as above takes place and the one-dimensional IDFT relation is:

$$E_m = \sum_{i=0}^{q-2} S_{q-1-i,0}\alpha^{mi} \tag{2.42}$$

where α is the primitive element of the finite field and E_m is the summation of all error values happening at the points of nonzero x-coordinate α^m. The property of Hermitian curves mentioned above still applies which says there are no points on the curve with the same x-coordinate value α^m and nonzero y-coordinate. Hence, E_m is in fact the error magnitude of the error at the point $P_{(x,0)} = (\alpha^m, 0)$.

Category four: A two-dimensional IDFT is used for errors occurring at the points of nonzero x-coordinate and nonzero y-coordinate, $P_{(x,y)}$. The error magnitude of the error at any point $P_i = (x, y)$ is given by:

$$e_i = \sum_{a=0}^{q-2}\sum_{b=0}^{q-2} S_{a,b} x_i^{-a} y_i^{-b} \tag{2.43}$$

where e_i is the error magnitude of the error that happened at the point P_i, and q is the size of the finite field. However, before Sakata et al. started this method in 1995, which was later improved by Liu [20] in 1999, a very lengthy and complex method was found by solving Eq. (2.17).

2.1.4 Complete Hard-Decision Decoding Algorithm for AG Codes Constructed From Hermitian Curves

In this section, we describe the details of the decoding algorithm used to decode AG codes constructed from Hermitian curves. It is used for iterative decoding [4, 5, 8] later in this book.

Step 1: Known syndromes computation:

a. The known syndromes $S_{0,0}, \ldots, S_{0,j}$ can be found by applying Eq. (2.17).
b. The known syndromes $S_{j+1,0}, \ldots, S_{m,j-m+1}$ can be found using Eq. (2.22).

Step 2: Finding the error location:
The known syndromes and some of the unknown syndromes up to $S_{0,j+m}$ are needed to find the error locations.

a. Run Sakata's algorithm with known syndromes (found in step 1) as input; some unknown syndromes are found using (2.22) when syndrome is of the form $S_{a,b}$ for $b \geq m - 1$.
b. Run Sakata's algorithm with unknown syndromes (found in step 2-a) as input; when having a syndrome of the form $S_{a,b}$ for $a \geq m$, then (2.22) is used to compute the value of the unknown syndrome or MV scheme is used if the syndrome has the form $S_{a,b}$ for $a < m$.

c. Run Sakata's algorithm with unknown syndromes (found step 2-b) as input and find more unknown syndromes using (2.22) when syndrome is of the form $S_{a,b}$ for $b \geq m - 1$.

d. Substitute the points on the curve into any of the minimal (error-locating) polynomials in set F to find its roots as these roots are the error locations.

Step 3: Finding magnitudes of errors:
The unknown syndromes from $S_{j+1+m,0}$ up to the last syndrome of the two-dimensional syndrome array $S_{q-1,q-1}$ are needed to compute the error magnitudes.

a. Equation (2.22) is used to find the value of the unknown syndrome if it is of the form $S_{a,b}$ for $a \geq m$.

b. If the unknown syndrome is of the form $S_{a,b}$ for $a < m$, then to compute its value, a recursive relationship between the syndromes should be formed by substituting the last minimal polynomial in the set F in Eq. (2.19).

c. Find the error values using IDFT:

- When the error location is at the origin point $P_{x,y} = (0, 0)$, then Eq. (2.39) is used to find the error magnitude.
- When the error is located at a point with zero x-coordinate and nonzero y-coordinate $P_{x,y} = (0, y)$, then Eq. (2.41) is used to find the error magnitude.
- When the error is located at a point with nonzero x-coordinate and zero y-coordinate $P_{x,y} = (x, 0)$, then Eq. (2.42) is used to find the error magnitude.
- When the error is located at a point with nonzero x-coordinate and nonzero y-coordinate $P_{x,y} = (x, y)$, then Eq. (2.43) is used to find the error magnitude.

Step 4: Error correction:
To correct the errors in terms of extracting the original message, the error values found in step 3 at the positions found in step 2 are added into the received codeword to give the decoded codeword. Then the original message is the first k symbols from the decoded codeword as the code is systematic.

2.2 Turbo Codes

Turbo coding was a breakthrough in channel coding introduced in 1993 by a group of French researchers [21, 22] as a new class of error correction codes with a relevant iterative decoding method. Turbo coding was not just a new set of codes but a new way of thinking about channel coding. These codes showed performance close to Shannon's capacity limit [21]. This represented a significant gain in power efficiency over other coding techniques known at that time.

The operation of a turbo codec relies on some basic ideas: using uncorrelated inputs, divide and conquer, and processing information iteratively. The information to be transmitted is stored in a memory in order to be scrambled to produce two

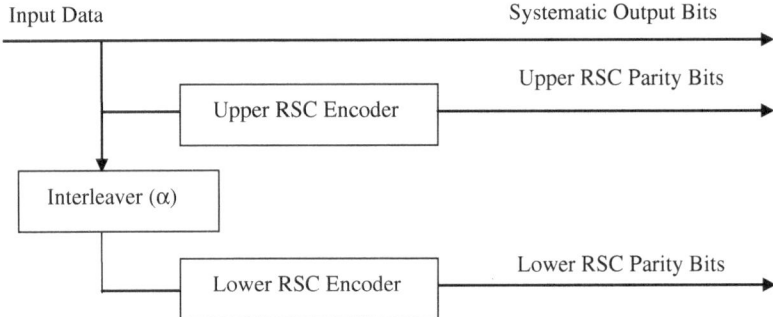

Input Data   Systematic Output Bits

Upper RSC Parity Bits

Upper RSC Encoder

Interleaver (α)

Lower RSC Parity Bits

Lower RSC Encoder

Fig. 2.4 Turbo encoder

uncorrelated sequences that are then encoded and transmitted. This idea is the key to the incomparable performance of turbo codes [23].

Since TCs were introduced, they have been useful for low-power applications such as satellite, deep-space communications, and for interference limited applications such as third generation (3G) cellular and personal communication services. Even though TCs have been a "hot topic" in the research literature over the past decade, there is still a relative lack of basic and fundamental papers serving as a starting point for researchers in this field [24]. The following sections in this chapter will briefly describe the main three components of a turbo codec (turbo encoder, interleaver, and turbo decoder).

2.2.1 Turbo Encoder

The basic turbo code encoder is produced using parallel concatenation of two identical recursive systematic convolutional (RSC) encoders separated by arbitrary interleaver (other interleavers could also be used such as block interleaver) [21, 25] as shown in Fig. 2.4.

This way of constructing an encoder is called parallel concatenation because the two encoders operate on the same input bits, rather than one encoding the output of the other. As a result, TCs are called parallel concatenated convolutional codes (PCCC) [26].

Both encoders have the same rate ($r = 1/2$), the upper encoder receiving data directly while the lower one receives it after being randomly interleaved by a permutation function α which maps bits in position i to position $\alpha(i)$. It is important to note that this interleaver α works in a block-wise manner, interleaving L bits at a time. Hence, TCs are actually block codes [25]. As both encoders receive the same input sequence in permuted fashion then only one of the systematic outputs needs to be transmitted. In most turbo encoders, the systematic output of the upper encoder is sent along with the parity bits of both of them. The overall rate of a

TC consisting of parallel concatenation of two systematic codes with rate ($r = 1/2$) is ($r = 1/3$). However, this rate can be increased if a subset of the parity bits is stopped from being transmitted by a process called puncturing. The code rate of a TC is increased to ($r = 1/2$) if the odd indexed parity bits and all systematic bits from the upper encoder are transmitted along with the even indexed parity bits from the other encoder [24].

2.2.2 Interleaver

The interleaver in turbo coding is a pseudorandom block scrambler which permutes N input bits with no repetitions by reading it into the interleaver and reading it out pseudorandomly [23, 25]. The interleaver has two main roles in TC: converting the small memory convolutional codes into long block codes, and decorrelating the inputs to both decoders so that an iterative sub-optimal decoding algorithm based on information exchange between the two decoders can be applied. This role of the interleaver makes it necessary that the same interleaving pattern should be available at the decoding side [21, 22]. If the input sequences to the two decoders are decorrelated, then there is a high possibility that after correction some of the errors in one decoder and some of the remaining errors become correctable in the second decoder [25].

2.2.3 Turbo Decoder

The TC decoder is constructed in a similar way as the encoder. Two simple soft-input soft-output (SISO) decoders are interconnected to each other in a serial concatenation. An interleaver is installed between the two decoders to spread out error bursts coming from the output of first decoder [21].

TCs can be decoded by maximum a posteriori (MAP) or maximum likelihood (ML) decoding methods. These decoders could be implemented only for small size interleavers as they are too complex for medium and large interleaver sizes [26]. The realistic value of TCs lies in the availability of a simple sub-optimal decoding algorithm [21, 26].

The idea behind turbo decoding is improving the reliability of the second decoder output by feeding it with extrinsic information that has been extracted out of the first decoder output. Then the reliability of the first decoder's output is improved by feeding the first decoder with extrinsic information extracted from the second decoder's output. This process will keep iterating until no further improvement can be made on the performance of the turbo decoder [24].

2.3 Block Turbo Codes

In 1994 Ramesh Pyndiah introduced BTCs as an alternative to classical convolutional TCs which were introduced a year before for applications requiring either high code rates ($R > 0.8$), very low error floors, or low complexity decoders that operate at several hundreds of megabits per second or higher [27].

BTCs are constructed as the data to be encoded is set in an l-dimensional hypercube with dimensional lengths denoted by ($k_1, k_2, \ldots, k_l$). Here all the dimensional sub-codes are encoded in the systematic linear block code (n_i, k_i, d_{mini}), where n_i represents the length of the code, k_i is the length of the information bit, d_{mini} is the minimum Hamming distance, and $r_i = k_i / n_i$ the code rate of the i-th dimensional sub-code. As a result for the l-dimensional BTC, the codeword length is $\prod_{i=1}^{l} n_i$; the information bit length is $\prod_{i=1}^{l} k_i$; the minimum Hamming distance is $\prod_{i=1}^{l} d_{mini}$; and the code rate is $\prod_{i=1}^{l} r_i$. Note that a higher dimensional number of the BTC implies a more complex implementation so the two-dimensional BTC seems to be the right choice for communication systems because of its relatively simple implementation and suitable structure for high code rate codes [28].

The RS code or Bose-Chaudhuri-Hocguenghem (BCH) code can be chosen as the component code of a two-dimensional BTC. The RS code has better error correction performance but due to its very high decoding complexity, the BCH code is usually preferred for practical applications [23].

To encode a two-dimensional BTC whose component code is a BCH code, first the $k_1 \times k_2$ information bits are set into a matrix of k_2 rows and k_1 columns. Then the k_2 rows are horizontally or row-wise encoded by applying $BCH(n_1, k_1, d_{min1})$ and k_1 columns are vertically or column-wise encoded by applying $BCH(n_2, k_2, d_{min2})$ as shown in Fig. 2.5 [25].

In addition, a row/column interleaver is used in between the two BCH encoders to guarantee the information bit that is horizontally encoded in the first BCH encoder can be vertically encoded in the second BCH encoder. One can see that this encoding technique is identical to encoding a BCH serial concatenated code in which the same interleaver used. Encoding with this technique leads to a BTC with the following parameters: $n = n_1 \times n_2$, $k = k_1 \times k_2$, and $d_{min} = d_{min1} \times d_{min2}$.

Concerning the decoding process, let us consider the decoding of binary linear block code $c(n, k, d_{min})$. While for high rate block code whose codeword length is too long, ML decoding requires very large code numbers and the complexity of the decoding algorithm increases exponentially. Therefore, a decoding technique with much lower complexity and small degradation in performance for the linear block code was introduced by Chase in 1972 and used by Pyndiah in 1994 [27]. It should be noted that the previous technique is also suitable for decoding non-binary codes like RS codes.

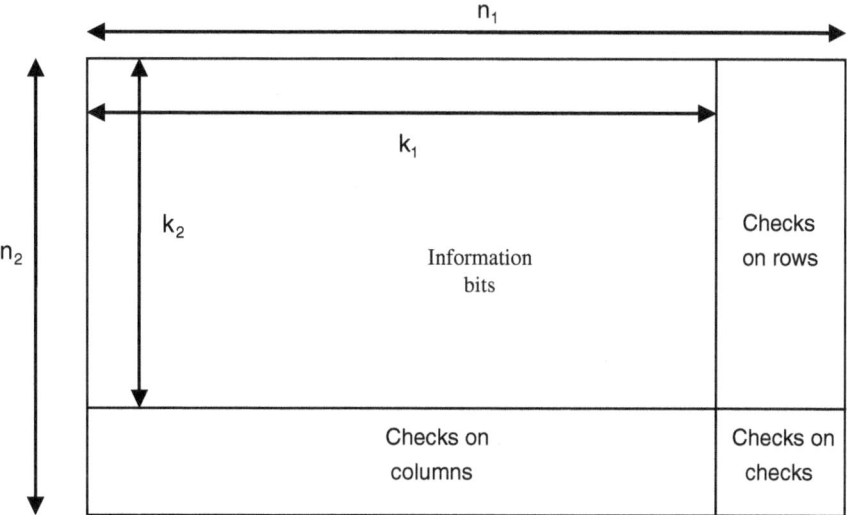

Fig. 2.5 The encoding structural diagram of the two-dimensional BTC

2.4 Summary

In this chapter, an overview of the concepts of AG code construction, encoding, and decoding techniques has been presented in detail which forms a foundation for understanding the subsequent chapters.

AG code construction sets out the code parameters such as the message length, codeword length, minimum Hamming distance, and the capabilities of code in locating and correcting errors. The encoding part was mainly composed of generating a non-systematic generator matrix and converting that into a systematic one using Gauss-Jordan elimination technique. For decoding, a full description was given of Sakata's algorithm and the MV technique was explained as well. Finding the magnitudes of errors depending on their locations was also explained.

Also in this chapter, the basics of TC were reviewed through a brief description of the main three components of TC (turbo encoder, interleaver, and turbo decoder). BTCs were introduced briefly as a prelude to more detailed explanations to follow in the next chapters.

References

1. Goppa VD (1981) Codes on algebraic curves. Soviet Math Dokl 24:75–91
2. Ozbudak F, Stichtenoth H (1999) Constructing codes from algebraic curves. IEEE Trans Inf Theory 45(7):2502–2505. doi:10.1109/18.796391
3. Tsfasman MA, Vladut SG, Zink T (1982) Modular curves, shimura curves and goppa codes, better than Varshamov-Gilbert bound. Math Nachrichten 109:21–28

4. Justesen J, Larsen K, Jensen H, Havemose A, Hoholdt T (1989) Construction and decoding of a class of algebraic geometry codes. IEEE Trans Inf Theory 35(4):811–821. doi:10.1109/18. 32157

5. Carrasco RA, Johnston M (2008) Non-binary error control coding for wireless communication and data storage. Wiley, New York. doi:10.1002/9780470740415.fmatter. http://dx.doi.org/10. 1002/9780470740415.fmatter

6. Kirwan F (1992) Complex algebraic curves. London mathematical society student texts, vol. 23. Cambridge University Press, Cambridge. doi:10.2277/0521423538

7. Walker RJ (1978) Algebraic curves. Princeton mathematical series, vol. 13. Springer-Verlag, New York

8. Johnston M, Carrasco RA (2005) Construction and performance of algebraic-geometric codes over awgn and fading channels. IEE Proc Commun 152(5):713–722. doi:10.1049/ip-com: 20045153

9. Blake I, Heegard C, Hoholdt T, Wei V (1998) Algebraic-geometry codes. IEEE Trans Inf Theory 44(6):2596–2618. doi:10.1109/18.720550

10. Pretzel O (1998) Codes and algebraic curves. Oxford University Press, New York

11. Wicker SB (1995) Error control systems for digital communication and storage. Prentice-Hall, Upper Saddle River

12. Sakata S (1988) Finding a minimal set of linear recurring relations capable of generating a given finite two-dimensional array. J Symbolic Comput 5(3):321–337. doi:10.1016/S0747-7171(88)80033-6. http://www.sciencedirect.com/science/article/pii/S0747717188800336

13. Atkinson K (1989) An introduction to numerical analysis, 2nd edn. Wiley, New York. http://www.amazon.com/exec/obidos/redirect?tag=citeulike07-20&path=ASIN/0471624896

14. Reed IS, Solomon G (1960) Polynomial codes over certain finite fields. J Soc Ind Appl Math 8(2):300–304

15. Massey J (1969) Shift-register synthesis and bch decoding. IEEE Trans Inf Theory 15(1):122–127. doi:10.1109/TIT.1969.1054260

16. Justesen J, Larsen K, Jensen H, Hoholdt T (1992) Fast decoding of codes from algebraic plane curves. IEEE Trans Inf Theory 38(1):111–119. doi:10.1109/18.108255

17. Sakata S, Justesen J, Madelung Y, Jensen H, Hoholdt T (1995) Fast decoding of algebraic-geometric codes up to the designed minimum distance. IEEE Trans Inf Theory 41(6):1672–1677. doi:10.1109/18.476240

18. Saints K, Heegard C (1995) Algebraic-geometric codes and multidimensional cyclic codes: a unified theory and algorithms for decoding using Grobner bases. IEEE Trans Inf Theory 41(6):1733–1751. doi:10.1109/18.476246

19. Feng GL, Rao TRN (1993) Decoding algebraic-geometric codes up to the designed minimum distance. IEEE Trans Inf Theory 39(1):37–45. doi:10.1109/18.179340

20. Liu CW (1999) Determination of error values for decoding Hermitian codes with the inverse affine Fourier transform. IEICE Trans Fundam Electron Commun Comput Sci 82(10):2302–2305. http://ci.nii.ac.jp/naid/110003208168/en/

21. Berrou C, Glavieux A, Thitimajshima P (1993) Near shannon limit error-correcting coding and decoding: turbo-codes. 1. In: IEEE International Conference on Communications (ICC'93), vol. 2. pp. 1064–1070. doi:10.1109/ICC.1993.397441

22. Berrou C, Glavieux A (1996) Near optimum error correcting coding and decoding: turbo-codes. IEEE Trans Commun 44(10):1261–1271. doi:10.1109/26.539767

23. Sklar B (1988) Digital communications: fundamentals and applications. Prentice-Hall, Upper Saddle River

24. Valenti MC (1998) Turbo codes and iterative processing. In: Proceedings IEEE New Zealand Wireless Communications Symposium, pp. 216–219

25. Vucetic B, Yuan J (2000) Turbo codes: principles and applications. Kluwer Academic, Boston. http://www.loc.gov/catdir/enhancements/fy0820/00033104-t.html

26. Benedetto S, Montorsi G (1996) Unveiling turbo codes: some results on parallel concatenated coding schemes. IEEE Trans Inf Theory 42(2):409–428. doi:10.1109/18.485713

27. Pyndiah R, Glavieux A, Picart A, Jacq S (1994) Near optimum decoding of product codes. In: Proceedings of IEEE GLOBECOM '94, pp. 339–343. doi:10.1109/GLOCOM.1994.513494
28. Pyndiah RM (1998) Near-optimum decoding of product codes: block turbo codes. IEEE Trans Commun 46(8):1003–1010. doi:10.1109/26.705396

Chapter 3
Literature Review

There is a relative lack of basic and fundamental chapters that can serve as a starting point for researchers in the field of using algebraic geometry theory in forward error correction and especially in BTCs. Even the algebraic geometry approach found to be efficient in dealing with binary and non-binary fields. So this chapter will concentrate on the construction and decoding aspects of AG codes to build up a sound knowledge to start developing the new BTC and IBTC.

The literature on constructing and decoding of AG codes is limited due to the fact that not many researchers are interested in working in this specialization as it requires a good knowledge and understanding of the theory of algebraic geometry, a difficult and complicated branch of mathematics.

3.1 Construction and Decoding of AG Codes

The existence of good linear codes were proven by Varshamov in 1957, showing that they have code rate $R = k/n$, minimum distance rate $\sigma = d/n$, and lower bounded by the Gilbert-Varshamov bound. The bound assures the existence of codes with longer and longer lengths but still with the same rate as probability of error goes to zero whereas the code length approaches infinity by using bounded distance decoding algorithms [1].

In 1981, Goppa [2] was the first to show the connection between the theory of algebraic geometry and error correcting codes, and showed an idea for efficient construction of very long codes with good parameters like relatively large minimum distance and high coding rate. This review of literature on AG codes will focus on the construction of "good" linear AG codes and the development of efficient decoding algorithms for AG codes.

No binary code having parameters exceeding this lower bound was known until a breakthrough made by Tsfasman, Vladut and Zink in 1984. Their work established that with very high complexity, it is possible to produce good linear AG codes

J. A. Alzubi et al., *Forward Error Correction Based On Algebraic-Geometric Theory*, 31
SpringerBriefs in Electrical and Computer Engineering,
DOI: 10.1007/978-3-319-08293-6_3, © The Author(s) 2014

exceeding the Gilbert-Varshamov bound using modular curves in certain square finite fields setting a new lower bound called Tsfasman-Vladut-Zink bound [3].

A new class of codes based on algebraic plane curves was introduced in 1989 by Justesen et al. [4]. Further they provided a detailed explanation of the process of constructing such codes (i.e., parameters, generator and parity-check matrices). They claimed that their method of construction is so simple that it does not require much knowledge in algebraic geometry theory. They also presented an algorithm for decoding that is considered to be a general form of Peterson's decoding algorithm for binary BCH codes and also a general form of the Peterson-Gorenstein-Zierler (PGZ) algorithm for short non-binary BCH and RS codes.

A modified version of this decoding algorithm was presented by Skorobogatov and Vladut in 1990 [5] to decode any AG code constructed from algebraic curve with errors correcting ability up to $[d^* - \gamma - 1]/2$ errors, where d^* is the designed minimum Hamming distance of the code and γ is the genus of the curve, with the same complexity as the PGZ algorithm. They also presented a version for the case of codes generated from elliptic and hyperelleptic curves with errors correcting ability of up to $[(d^* - 1)/2]$ errors.

Later in 1992, Justesen et al. [6] used Sakata's algorithm from 1988 [7] to reduce the complexity of his famous decoding algorithm described earlier.

Sakata's 1988 algorithm [7] was able to find a minimal set of two-dimensional linear recurring relations to generate a two-dimensional array containing syndromes from which a set of minimal polynomials is generated. The coefficients of these minimal polynomials will form a recursive relationship between the syndromes in the two-dimensional array. The errors locations can be found by finding the points on the curve that make any of these minimal polynomials vanish.

It is worth mentioning that Sakata demonstrated how higher dimensions can be achieved through extensions of his algorithm. The worst-case computation for syndrome array of size n is $O(n^2)$. However, the overall computational complexity of Sakata's algorithm has worst-case of $O(n^7/3)$. This process is a two-dimensional extension of the BM algorithm [8, 9] which uses a one-dimensional vector of syndromes to generate a minimal polynomial. A recursive relationship between these syndromes is created from the coefficients of the polynomial. The locations of errors can be found by inverting the roots of the minimal polynomial. The BM algorithm has a worst-case computation complexity of $O(n^2)$ which is better than Sakata's algorithm.

In 1993, Feng and Rao [10] introduced a simple MV scheme in order to overcome the shortcoming of all previous decoding algorithms which can be summarized as an inability of these algorithms to correct a number of errors up to the maximum number that can be achieved by the algorithm. The purpose of Feng and Rao's work was to simplify the concept of AG codes and introduce a decoding algorithm. The idea of their algorithm was to apply Gaussian elimination on a matrix of known syndromes and use the MV scheme in order to find the values of unknown syndromes. Having extra syndromes enables this algorithm to correct $\lfloor d^* - 1/2 \rfloor$ errors. Basically, this decoding algorithm was a generalization of Peterson's decoding algorithm for BCH codes with computation complexity of $O(n^3)$.

Duursma applied the MV scheme [11] to Skorobogatov and Vladut's procedure introduced in 1990 in order to increase the algorithm's error correction capability. However, the computational complexity remained the same as in [5] which is $O(n^3)$.

Another simple method for constructing AG codes generated from affine plan curves was proposed by Feng et al. in 1994 [12]. They introduced a fast decoding technique with complexity less than the first decoding algorithm presented by Justesen et al. in 1989 and the decoding algorithm of Skorobogatov and Vladut as well, with capability to correct up to $[(d^* - 1)/2]$ errors.

A year later [13], the same authors [12] presented a simple construction method for AG codes from algebraic curves and other varieties with better parameters than traditional AG codes when high code rate and large genus are considered.

Construction of codes from elliptic curves been studied by Yaghoobian and Blake [14]. Elliptic curves produce maximal curves with property of having maximum number of points for different finite fields of characteristic two. Sakata with Justesen et al. in 1995 [15] introduced the MV scheme of Feng and Rao to Sakata's algorithm [16] which is a generalization of the BM algorithm. They were able to correct all errors of weight less than $d^*/2$ with low computational complexity $O(n^{7/3})$. The only restriction for this algorithm is that it is not able to correct any errors occurring at any point with a zero coordinate.

In contrast to all previous construction methods which produce non-systematic codes; Heegard et al. [17] in 1995 were able to present the first systematic AG codes based on the theory of Grobner bases which provides a description and implementation of a systematic encoder.

Later in 1998, Blake et al. [18] developed AG codes from particular classes of curves, e.g., elliptic, hyperelliptic, and Hermitian curves. They also presented decoding algorithms for these classes of curve codes.

In 1999, Xing et al. [19] introduced two construction methods for linear codes from local expansions of functions at a fixed rational point. While their constructions have the same bound on the parameters as Goppa's codes and equivalent to Goppa's construction method, the codes they constructed from maximal curves turned out to have better parameters than the codes obtained by Goppa from maximal curves with the restriction of a certain interval of parameters.

The problem of correcting errors that are located at points with a zero coordinate, which was considered a drawback of the modified version of Sakata's decoding algorithm, was addressed in [15]. It was resolved later in 1999 by Liu [20].

There were no simulation results evaluating the performance of AG codes with hard-decision decoding algorithms until 2004, when Johnston et al. [21] introduced their first simulation results for designing AG codes over fading channels using a BPSK modulation scheme. As they stated [21], AG codes have the property of longer code lengths compared to RS codes. In addition, there are more choices of codes with acceptable decoding complexity. Significant coding gains over fading channels have been demonstrated in simulation results of AG codes and RS codes, maintaining the same code rate and same finite field but not the code size.

A year later, Johnston et al. [22] presented a simulation work of systematic AG codes constructed from Hermitian curves (Hermitian codes) over additive white

Gaussian noise (AWGN) and Rayleigh fast fading channels using BPSK modulation. They showed again that AG codes outperformed RS codes and suggested a possible future use of these codes in many fields such as mobile radio environment where RS codes are not suitable because of their length and the limitation on their number.

However, all previous simulation results were concerned with evaluating and comparing AG codes performance with linear block codes such as RS codes. In other words, no evaluation and comparison of BTCs using AG codes as code components has been done. In addition to that, simulation of performance for IBTCs using AG codes as code components does not exist in the current literature.

3.2 Iterative Decoding of Block Turbo Codes

The need for high code rates ($R > 0.8$), very low error floors, and low-complexity decoders that operate at high rate have been driven by the adoption of real-time data services such as video transmission and other real-time video applications. These applications led to the introduction of TCs and iterative decoding by Berrou et al. for the first time in 1993 [23].

The early implementation of TCs was in satellite and deep-space missions in which they showed impressive BER performance compared to the codes being used at that time without requiring additional power. Due to this property they played an important role in many commercial applications such as third generation (3G) wireless phones, Digital Video Broadcasting (DVB) systems, or wireless metropolitan area networks (WMAN), etc.

In 1994, Ramesh Pyndiah et al. extended the idea to BTC or what is known as Turbo Product Codes (TPCs) achieved by serially concatenating two block codes [24]. Later these codes were viewed as an attractive alternative choice to the classical convolutional turbo codes (CTCs). Pyndiah et al. introduced a new decoding scheme known as "Chase-Pyndiah" soft decoder to improve the BER performance of the block codes as hard-decision decoding algorithms in use before that. The main and important idea that the turbo decoding relies on is the exchange of probabilistic messages (extrinsic information) between the SISO decoders.

Since the introduction of the Chase-Pyndiah SISO decoding algorithm in 1994, continuous improvements have been made by researchers with the aim to lower decoding complexity, improve BER performance, and increase coding gain. In 1999, Picart and Pyndiah [25] claimed that a coding gain of up to 2 dB can be achieved in short codes, and a reduction by 1 or 2 decoding steps can be achieved as well for specific BER in long codes. The results were obtained by adaptation of decoding algorithm to the characteristics of the encoder, modulation, and the number of decoding steps.

Later in 2001, Hirst et al. [26] introduced a highly efficient fast Chase decoding algorithm by reordering the original Chase algorithm's repeated decodings such that the inherent computational redundancy is greatly reduced without any reduction in performance.

A year later, Martin and Taylor suggested an alternative Chase-based decoding algorithm for BTCs [27]. The idea of the proposed algorithm is to calculate the distance of only a small subset of codewords in order to estimate the extrinsic information.

Non-binary BTCs have been of interest to researchers due to their large minimum Hamming distances, less sensitivity to puncturing patterns, reduced latency, robustness of the decoder, and better convergence [28]. The iterative decoding and performance of non-binary BTCs have been improved since the introduction of turbo decoding.

The year 1996 witnessed the first appearance of non-binary product codes when Aitsab and Pyndiah [29] introduced the iterative decoding of RS product codes. They presented two construction methods for these codes, and showed that the iterative decoding of this new coding scheme is based on the soft decoding and the soft decision of the component codes. The evaluation of the performance of this class of codes over an AWGN channel showed a coding gain of up to 5.5 dB for BER 10^{-5}. The achieved results made these codes very attractive for data storage applications.

Later in 2000, Sweeney and Wesemeyer [30] claimed that a very good coding gain in terms of BER performance and a reduction in complexity can be obtained when using the sub-optimal soft-decision Dorsch's algorithm combined with Pyndian's method for extracting soft output to iteratively decode block codes defined over finite fields higher than $GF(2)$. Their chapter presented two new different interleaving structures which yield different performances in terms of coding delay and BER performance.

Zhou et al. in 2004 [31] presented a comparison between BTCs constructed based on Q-ary symbol concatenation and BTCs constructed based on bit concatenation of about similar coding rates. They showed that the aforementioned class of codes outperforms the latter in terms of BER performance with lower hardware complexity [31]. They also claimed that the Q-ary symbol based concatenation BTCs can achieve reliable transmission at less than one dB away from Shannon's bound, when proper choice of component codes is made. For high code rate applications such as high speed optical transmissions and data storage, the authors in [31, 32] found that the Q-ary symbol concatenation BTCs are more suitable than the ones constructed based on bit concatenation as they have much smaller data block size which is directly proportional to the coding/decoding delay and size of memory in use.

In the same year, Diatta et al. [33] showed clearly that the turbo RS iterative decoding based on Pyndiah's method of extracting the soft output used in enhanced very high bit rate digital subscriber line (VDSL) systems perform much better than the classical RS hard decision decoding used in asymmetric digital subscriber line (ADSL).

In 2006, Piriou et al. [34] introduced an efficient non-binary BTC decoder architecture. The authors presented an architecture for BTCs using RS codes as component codes, in which they implemented the key equation solver for the algebraic decoding of RS codes which is considered a design innovation in this architecture. Another design innovation was reported implementing the iterative SISO decoding of RS-BTCs in this efficient architecture. Building the architecture this way makes this family of codes more suitable for applications requiring high code rates such as mobile communications, data storage, satellite communications, and optical communications. It is worth noting that this architecture was the first published architecture to implement a RS-BTC decoder.

3.3 Irregular Iterative Decoding of Block Turbo Codes

In recent years, great interest has been shown in the concept of unequal protection of information bits. The concept has been deployed in the design of irregular low density parity check (LDPC) codes, irregular turbo codes (ITCs), and BTCs. The idea is attractive because of these advantages: improved BER performance, reduced decoding complexity compared to the regular (equal protection) codes, and codes that are close to Shannon's bound.

In 1999, the first ITC was presented in a chapter titled "Irregular Turbo Codes" by Frey and MacKay [35]. The authors claimed that a coding gain of 0.15 dB is obtainable at BER 10^{-4} over an AWGN channel using BPSK modulation. It was accomplished by changing the structure of the original rate $1/2$ TC of Berrou et al. to be slightly irregular. They also showed that the BER performance of this new irregular TC performs in the same regime as the best known irregular Gallager code at that time.

A year later the same team [36] showed that an increase in the rate of codes that compose an irregular code will cause the number of low-weight codewords to be increased, which in turn produce an ITC. They further explained that it is possible to use the sum-product decoding algorithm—a general form of the turbo coding algorithm with low complexity—iteratively to decode their ITC. Their work [35, 36] suffered from a requirement for large frame size, though no report is available on what number of iterations is required to obtain a low BER performance.

Richardson et al. [37] showed the best irregular LDPC code with a length of one million bits which performed close to Shannon's bound over a noisy Gaussian channel using BPSK modulation. This new code showed an improvement in the performance of the LDPC codes of about 0.82 dB and was 0.13 dB away from Shannon's capacity at BER 10^{-6}. However, the cost for this improvement in performance was more complexity in the decoding process. The just mentioned result, the design, and the construction method were presented in their chapter titled "Design of Capacity-Approaching Irregular Low-Density Parity-Check Codes" published in 2001.

In 2003, Sawaya and Boutros [38] introduced an ITC design to lower the decoding complexity which is the point to be considered when applying channel coding. Their design consisted of a single RSC encoder and a single SISO decoder. However, it has two drawbacks: in order to achieve a very low bit error rate (i.e., 10^{-6}), it required a high number of iterations (nearly 100) and a very large frame size. The proposed design in [38] showed a coding gain of about 0.24 dB at BER of 10^{-6} over an AWGN channel using a BPSK modulation scheme compared to the regular TC.

As the complexity in all ITCs was the price for improving the BER performance and getting closer to Shannon's bound, Sholiyi in his thesis titled "Irregular Block Turbo Codes for communication systems" [39] introduced a lower complexity irregular block turbo codec for communication systems over noisy Gaussian channels which is flexible and high speed. The BER performance improved in these new codecs as they benefit from extra protection of some bits set in a specific manner using state of art techniques. The simulation results presented in [39] showed that IBTC having more coding gain over noisy Gaussian channels using higher modulation schemes (i.e., 16 QAM and 64 QAM modulation schemes) comparing to existing BTCs.

3.4 Summary

This literature review has discussed various methods for constructing and decoding AG channel codes. The performance of different AG codes in terms of BER were studied and evaluated in comparison with decoding complexity over AWGN and Rayleigh fast fading channels. The benefits and drawbacks of each method were highlighted. The simplest construction method was identified in order to use it in constructing the AG codes which are the focus of this book. The best decoding algorithm for AG codes in terms of complexity and BER performance was considered as well.

Interestingly, we have found in the literature that no one has considered AG codes as component codes for binary and non-binary TCs and BTCs. Also in this chapter, the techniques for regular decoding of BTCs were highlighted in order to use it in conjunction with the decoding algorithm of AG codes as component decoders of BTC decoders.

The irregular decoding methods of BTCs were studied and their benefits and drawbacks were highlighted in this chapter as BTCs will be revisited later in this book. We also found that irregular decoding has never been used in AG-BTCs, so the construction of irregular decoders is studied later.

References

1. Berlekamp ER (1972) A survey of algebraic coding theory: lectures held at the department for automation and information. In: Courses and lectures—international centre for mechanical sciences. Springer, Berlin. http://books.google.co.uk/books?id=i-RQAAAAMAAJ
2. Goppa VD (1981) Codes on algebraic curves. Soviet Math Dokl 24:75–91
3. Tsfasman MA, Vladut SG, Zink T (1982) Modular curves, Shimura curves and Goppa codes, better than Varshamov-Gilbert bound. Math Nachtrichten 109:21–28
4. Justesen J, Larsen K, Jensen H, Havemose A, Hoholdt T (1989) Construction and decoding of a class of algebraic geometry codes. IEEE Trans Inf Theory 35(4):811–821. doi:10.1109/18.32157
5. Skorobogatov AN, Vladut SG (1990) On the decoding of algebraic-geometric codes. IEEE Trans Inf Theory 36(5):1051–1060. doi:10.1109/18.57204

6. Justesen J, Larsen K, Jensen H, Hoholdt T (1992) Fast decoding of codes from algebraic plane curves. IEEE Trans Inf Theory 38(1):111–119. doi:10.1109/18.108255

7. Sakata S (1988) Finding a minimal set of linear recurring relations capable of generating a given finite two-dimensional array. J Symbolic Comput 5(3):321–337. doi:10.1016/S0747-7171(88)80033-6. http://www.sciencedirect.com/science/article/pii/S0747717188800336

8. Massey J (1969) Shift-register synthesis and bch decoding. IEEE Trans Inf Theory 15(1):122–127. doi:10.1109/TIT.1969.1054260

9. Berlekamp ER (1984) Algebraic coding theory. No. M-6. Aegean Park Press, California. http://books.google.co.uk/books?id=leSbQgAACAAJ

10. Feng GL, Rao TRN (1993) Decoding algebraic-geometric codes up to the designed minimum distance. IEEE Trans Inf Theory 39(1):37–45. doi:10.1109/18.179340

11. Duursma IM (1993) Majority coset decoding. IEEE Trans Inf Theory 39(3):1067–1070. doi:10.1109/18.256518

12. Feng GL, Rao TRN (1994) A simple approach for construction of algebraic-geometric codes from affine plane curves. IEEE Trans Inf Theory 40(4):1003–1012. doi:10.1109/18.335972

13. Feng GL, Rao TRN (1995) Improved geometric Goppa codes. i. basic theory. IEEE Trans Inf Theory 41(6):1678–1693. doi:10.1109/18.476241

14. Yaghoobian T, Blake I (1994) Reed-solomon codes and their applications, Chap. 13. IEEE Press, Piscataway, USA, pp 293–314

15. Sakata S, Justesen J, Madelung Y, Jensen H, Hoholdt T (1995) Fast decoding of algebraic-geometric codes up to the designed minimum distance. IEEE Trans Inf Theory 41(6):1672–1677. doi:10.1109/18.476240

16. Sakata S (1990) Extension of the Berlekamp-Massey algorithm to N dimensions. Inf Comput 84(2):207–239. doi:10.1016/0890-5401(90)90039-K. http://dx.doi.org/10.1016/0890-5401(90)90039-K

17. Heegard C, Little J, Saints K (1995) Systematic encoding via Grobner bases for a class of algebraic-geometric Goppa codes. IEEE Trans Inf Theory 41(6):1752–1761. doi:10.1109/18.476247

18. Blake I, Heegard C, Hoholdt T, Wei V (1998) Algebraic-geometry codes. IEEE Trans Inf Theory 44(6):2596–2618. doi:10.1109/18.720550

19. Xing C, Niederreiter H, Lam KY (1999) Constructions of algebraic-geometry codes. IEEE Trans Inf Theory 45(4):1186–1193. doi:10.1109/18.761259

20. Liu CW (1999) Determination of error values for decoding Hermitian codes with the inverse affine Fourier transform. IEICE Trans Fundam Electron Commun Comput Sci 82(10):2302–2305. http://ci.nii.ac.jp/naid/110003208168/en/

21. Johnston M, Carrasco R, Burrows BL (2004) Design of algebraic-geometric codes over fading channels. Electron Lett 40(21):1355–1356. doi:10.1049/el:20045392

22. Johnston M, Carrasco RA (2005) Construction and performance of algebraic-geometric codes over awgn and fading channels. IEE Proc Commun 152(5):713–722. doi:10.1049/ip-com:20045153

23. Berrou C, Glavieux A, Thitimajshima P (1993) Near shannon limit error-correcting coding and decoding: Turbo-codes. 1. In: IEEE ICC'93, vol 2, pp 1064–1070. doi:10.1109/ICC.1993.397441

24. Pyndiah R, Glavieux A, Picart A, Jacq S (1994) Near optimum decoding of product codes. In: IEEE GLOBECOM '94, pp 339–343. doi:10.1109/GLOCOM.1994.513494

25. Picart A, Pyndiah R (1999) Adapted iterative decoding of product codes. In: IEEE GLOBE-COM '99, vol 5, pp 2357–2362. doi:10.1109/GLOCOM.1999.831724

26. Hirst SA, Honary B, Markarian G (2001) Fast chase algorithm with an application in turbo decoding. IEEE Trans Commun 49(10):1693–1699. doi:10.1109/26.957387

27. Martin P, Taylor D (2002) Distance based adaptive scaling in suboptimal iterative decoding. IEEE Trans Commun 50(6):869–871. doi:10.1109/TCOMM.2002.1010602

28. Berrou C, Jezequel M, Douillard C, Kerouedan S (2001) The advantages of non-binary turbo codes. In: 2001 IEEE information theory workshop, pp 61–63. doi:10.1109/ITW.2001.955136

29. Aitsab O, Pyndiah R (1996) Performance of reed-solomon block turbo code. In: IEEE GLOBE-COM '96, vol 1, pp 121–125. doi:10.1109/GLOCOM.1996.594345
30. Sweeney P, Wesemeyer S (2000) Iterative soft-decision decoding of linear block codes. IEE Proc Commun 147(3):133–136. doi:10.1049/ip-com:20000300
31. Zhou R, Picart A, Pyndiah R, Goalie A (2004a) Reliable transmission with low complexity reed-solomon block turbo codes. In: 1st international symposium on wireless communication systems, pp 193–197. doi:10.1109/ISWCS.2004.1407236
32. Zhou R, Picart A, Pyndiah R, Goalic A (2004b) Potential applications of low complexity non-binary high code rate block turbo codes. In: IEEE MILCOM 2004, vol 3, pp 1694–1699. doi:10.1109/MILCOM.2004.1495192
33. Diatta De Geest D, Geller B (2004) Reed-solomon turbo codes for high data rate transmission. In: IEEE 59th VTC 2004, vol 2, pp 1023–1027. doi:10.1109/VETECS.2004.1388986
34. Piriou E, Jego C, Adde P, Le Bidan R, Jezequel M (2006) Efficient architecture for reed-solomon block turbo code. In: IEEE ISCAS 2006, p 4. doi:10.1109/ISCAS.2006.1693426
35. Frey B, Mackay D (1999) Irregular turbo codes. In: 37th allerton conference on communication, control and computing. Allerton House, Illinois
36. Frey B, Mackay D (2000) Irregular turbo-like codes. In: 2nd international symposium on turbo codes and related topics. Brest, France, pp 67–72
37. Richardson TJ, Shokrollahi MA, Urbanke RL (2001) Design of capacity-approaching irregular low-density parity-check codes. IEEE Trans Inf Theory 47(2):619–637. doi:10.1109/18.910578
38. Sawaya HE, Boutros JJ (2003) Irregular turbo codes with symbol-based iterative decoding. In: 3rd international symposium on turbo codes and related topics. Brest, France
39. Sholiyi A (2011) Irregular block turbo codes for communication systems. Ph.D. thesis, Swansea University, Swansea, UK

Chapter 4
Algebraic-Geometric Non-binary Block Turbo Codes

In Chap. 2, the necessary mathematics needed to understand the design, construction, and encoding and decoding of AG codes were covered. This chapter will focus on the concept of block turbo design of AG codes constructed from Hermitian curves defined over finite fields, and the iterative decoding of the constructed block turbo codes using a HIHO decoding technique based on Sakata's algorithm with MV technique and Chase-Pyndiah's algorithm to extract a soft output from the hard output of the AG decoder. Then this chapter will present simulation results for BER performance of AG-BTCs compared with the BER performance of RS-BTCs of about same size and relatively similar rate over different finite fields.

4.1 AG Non-binary Block Turbo Code Encoder

The AG block turbo encoder consists of two AG systematic encoders (recall Chap. 2) separated by a block interleaver. The construction of an AG non-binary BTC is similar to the construction of the binary BCH-BTC except that each non-binary symbol consists of m bits. In other words, the AG non-binary BTC operates in a Galois field m for various AG non-binary BTC sizes [1, 2] which means that these codes consist of $K_1 \times K_2 \times q$ information bits where K_1, K_2 are shown in Fig. 2.5, and q is the Q-ary of the non-binary symbol.

The information symbols are arranged in a $k \times k$ block and encoded horizontally or row-wise by the first AG systematic encoder (outer AG systematic encoder) as illustrated in Fig. 4.1. The output from the previous step will be passed to the interleaver (inverting rows into columns and vice versa). The result will be fed to the second AG systematic encoder (inner AG encoder) which in a real sense means encoding the information symbols vertically or column-wise. It should be noted that the term "interleaving" does not exist in the BTC literature as the vertical encoding of the information symbols proceeding the horizontal encoding is the same as a block interleaved version of the horizontal symbols. Figure 2.5 shows structural diagram of the two-dimensional BTC.

J. A. Alzubi et al., *Forward Error Correction Based On Algebraic-Geometric Theory*,   41
SpringerBriefs in Electrical and Computer Engineering,
DOI: 10.1007/978-3-319-08293-6_4, © The Author(s) 2014

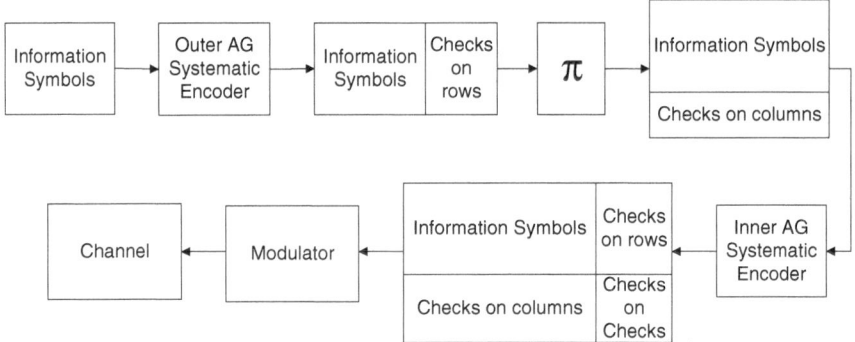

Fig. 4.1 AG non-binary BTC encoder

The output of the inner AG systematic encoder is a $n \times n$ symbols block which will be converted into bits to be modulated by applying any the following modulation schemes: BPSK, QPSK, 16-QAM, or 64-QAM. The modulated bits will be passed to an AWGN or Rayleigh fast fading channel.

4.2 AG Non-binary Block Turbo Code Decoder

To the best of our knowledge, the AG block turbo decoder shown in Fig. 4.2 is the first appearance in the literature. It consists of two AG decoders, a block interleaver, and a deinterleaver. The decoding is performed at symbol level as the AG decoder is a HIHO decoder. Iterative decoding is applied to AG codes to enhance its performance. This is done with the use of Chase-Pyndiah's decoding algorithm which consists of two main parts, a soft-input hard-output (SIHO) decoding algorithm and a hard-output computation unit, which is similar to the one implemented by Pyndiah in 1996 [3] for extracting soft output from a hard decision decoder.

The received sequence from the channel is demodulated and the soft information represented by a row or a column $E = (e_{1,1}, e_{1,2}, \ldots, e_{n,q})$, where n is the codeword length in symbols and q is the number of bits per symbol. The received codeword R is denoted by $R = [r_{1,1}, r_{1,2}, \ldots, r_{n,q}]$ where n is the codeword length in symbols and q is the number of bits per symbol.

The log-likelihood ratio (LLR) of each bit in R is computed using the general expression depending on the modulation scheme used [4]:

$$LLR\left(e_{i,j}\right) = \ln \frac{Pr\left\{e_{i,j} = +1/r_{i,j}\right\}}{Pr\left\{e_{i,j} = -1/r_{i,j}\right\}} \tag{4.1}$$

where $(i = 1, 2, \ldots, n)$ represents the codeword length in symbols and $(j = 1, 2, \ldots, q)$ represents the number of bits in each symbol.

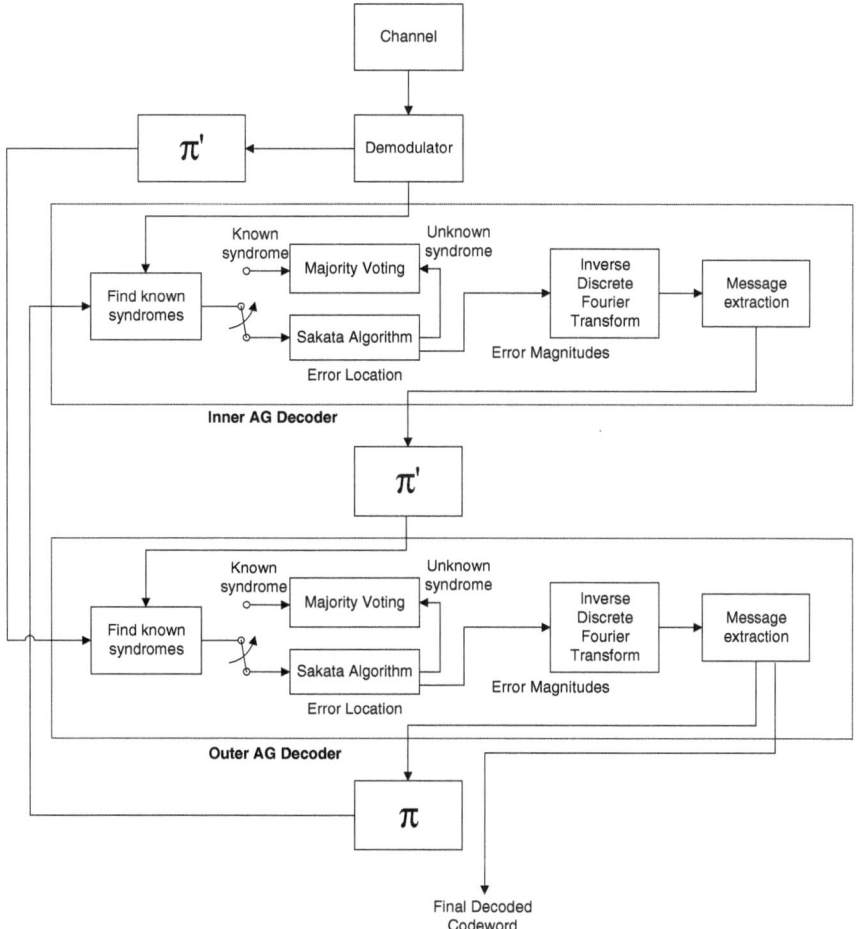

Fig. 4.2 AG non-binary BTC decoder

The hard decision Y^o of the transmitted signal is calculated using the sign of the LLR values for each received bit:

$$Y^o = [y_{11} \quad y_{12} \quad y_{13} \quad \cdots \quad y_{1q} \quad y_{21} \quad y_{22} \quad \cdots \quad y_{nq}] \tag{4.2}$$

and

$$Y^o = \begin{cases} +1 & \text{if } LLR(e_{i,j}) \geq 0 \\ -1 & \text{if } LLR(e_{i,j}) < 0 \end{cases} \tag{4.3}$$

This hard form of the received sequence is then passed through Chase-Pyndiah's algorithm, which is explained next in this chapter. Then the testing patterns

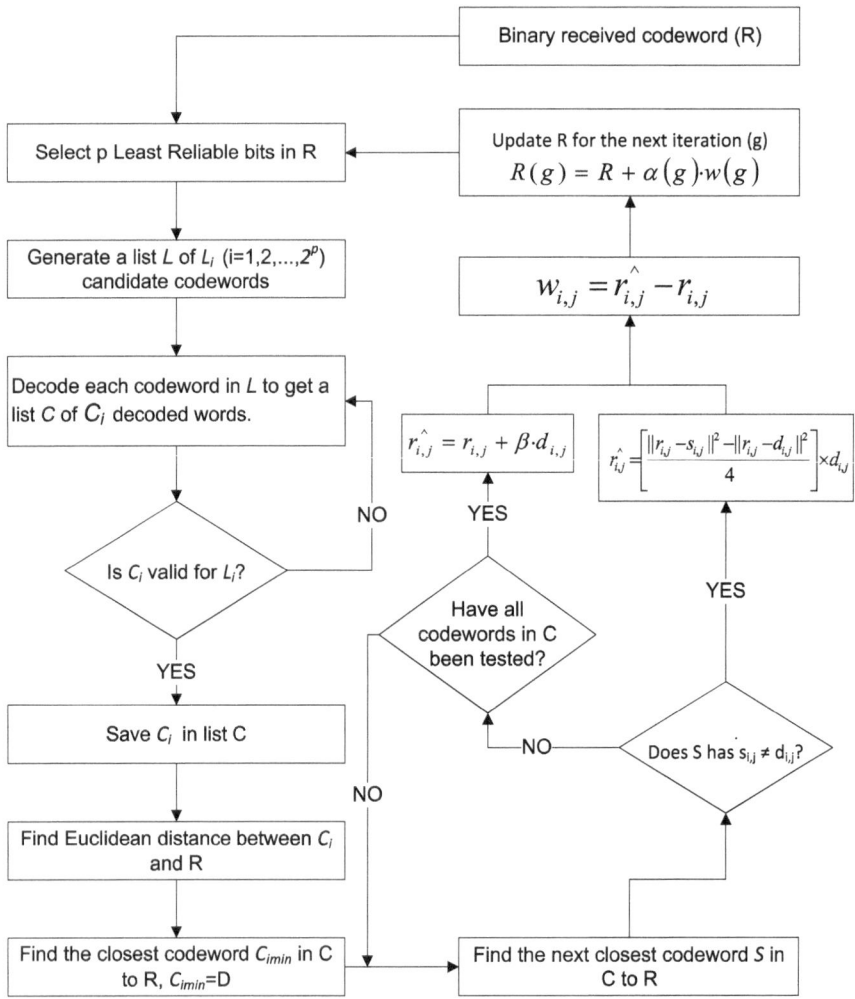

Fig. 4.3 Binary Chase-Pyndiah Algorithm

(2^p candidate codewords) are decoded using the inner AG decoder after being converted from binary symbols into non-binary symbols. The extrinsic information $W_{i,j}$ is then computed. The received sequence is updated by adding the extrinsic information into it to go into the second Chase-Pyndiah's algorithm, and next to the outer AG decoder after being deinterleaved using the block deinterleaver in the other half of the iteration.

4.2.1 Extracting Soft Information From the Hard Output of AG Decoder Using Chase-Pyndiah Algorithm

Chase-Pyndiah's algorithm is explained to take advantage of increasing the viewing range of the decoder, which is the common way of decoding BTCs for both binary and non-binary cases. It is illustrated in Fig. 4.3 [4].

It finds the p least reliable binary symbols in R and masks them (flipping $+1$ to -1 and vice versa) to obtain a list L containing 2^p candidate codewords (test patterns) denoted by L_i where $i = 1, 2, \ldots, 2^p$. The binary candidatecodewords are then converted into non-binary candidate codewords and decoded using the AG decoder based on Sakata's algorithm explained earlier in Chap. 2.

The decoding result of each candidate codeword in the list L is converted into binary symbols and stored in a list C which will contain at most 2^p distinct candidate codewords as there might be some repeated codewords in the set. The minimum Euclidean distance metric criterion is considered to find the nearest test pattern (candidate codeword) C_{imin} in C to the received word R which will be the final hard decision as in the following equation:

$$D = C_{imin} \quad \text{if} \quad \left| r - C^d \right| < \left| r - C^e \right| \quad \forall \ d \neq e \tag{4.4}$$

where D represents the selected and final hard decision codeword, and C_d and C_e are different candidate codewords in the list C. Let $D = [d_{1,1}, d_{1,2}, \ldots, d_{n,q}]$ where n is the codeword length in symbols, and q is the number of bits per symbol.

The next step is central to the turbo concept, which is extracting the extrinsic information from the selected candidate codeword D to update the soft input of the following iteration. In order to achieve this, the reliability of each bit in D based on R is computed using LLR as in the following equation which was illustrated in [5]:

$$LLR_{i,j} = \ln \frac{Pr\left\{e_{i,j} = +1/R\right\}}{Pr\left\{e_{i,j} = -1/R\right\}} \tag{4.5}$$

Applying normalisation, expansion, and approximation will yield:

$$r'_{i,j} = \frac{\sigma^2}{2}LLR_{i,j} = r_{i,j} + W_{i,j} \tag{4.6}$$

where $W_{i,j}$ is the required extrinsic information needed to update the soft input to the following iteration, and $r'_{i,j}$ is the soft output of the bit $d_{i,j}$ in the candidate codeword D which is calculated using the following equation [6]:

$$r'_{i,j} = \frac{\|r_{i,j} - C_{i,j}\|^2 - \|r_{i,j} - d_{i,j}\|^2}{4} \times d_{i,j} \tag{4.7}$$

where C is the next closest codeword to R in the list C having the bit $c_{i,j}$ with a minimum Euclidean distance from the bit $r_{i,j}$ such that $c_{i,j} \neq d_{i,j}$, and $d_{i,j}$ is the hard decision for each bit of the selected codeword. In this notation, $\| \ \|^2$ represents the norm. If the next closest codeword in C to R cannot be found, then the soft output from the bit $d_{i,j}$ in the selected candidate codeword D is $r'_{i,j}$ which can be calculated and defined using the following equation:

$$r'_{i,j} = r_{i,j} + \beta \cdot d_{i,j} \tag{4.8}$$

Here $r_{i,j}$ represents the j-th bit in the i-th non-binary received symbol, and β is a weighting factor that can be set as a constant ranging between 0 and 1 which increases as the iteration increases or approximated as in the following LLR [7]:

$$\beta \approx \ln \frac{Pr\left\{d_{i,j} = e_{i,j}\right\}}{Pr\left\{d_{i,j} \neq e_{i,j}\right\}} \tag{4.9}$$

The values of β used here are in the range between 0.2 and 0.85 at intervals of 0.1. However, in the decoding of the AG-BTC, the horizontal decoder consists of one Chase-Pyndiah decoding process which is a half iteration, while the horizontal and vertical decoders contain two Chase-Pyndiah decoding processes which are a full iteration. Each Chase-Pyndiah decoding process uses one β value. This implies that a full iteration requires two β values.

After computing $r'_{i,j}$, $W_{i,j}$ for each binary element in the codeword is computed using Eq. (4.6). The next step is to update the elements of R following:

$$r_{i,j}(g) = r_{i,j} + \alpha(g) \cdot W_{i,j}(g) \tag{4.10}$$

where g represents the number of the next decoding iteration, and α is a scaling factor that reduces the influence of extrinsic information delivered at the previous half iteration. The α values used here increase with the iteration number and range between 0 and 0.7 at intervals of 0.1.

Several Systematic AG-BTCs constructed from Hermitian curves over $GF(2^4)$ were evaluated in terms of their performance using Monte Carlo simulations and compared with RS-BTCs codes over $GF(2^8)$ of about same size and similar code rate. The simulation results showed that this coding scheme outperforms comparable RS schemes over both AWGN and Rayleigh fast fading channels.

4.3 BER Performance of AG Block Turbo Codes Versus RS Block Turbo Codes

In this section, the performance of AG-BTCs is compared to performance of RS-BTCs. The process of making AG codes function in an iterative manner is carried out considering the number of iterations, number of least reliable (LR) bits, and code

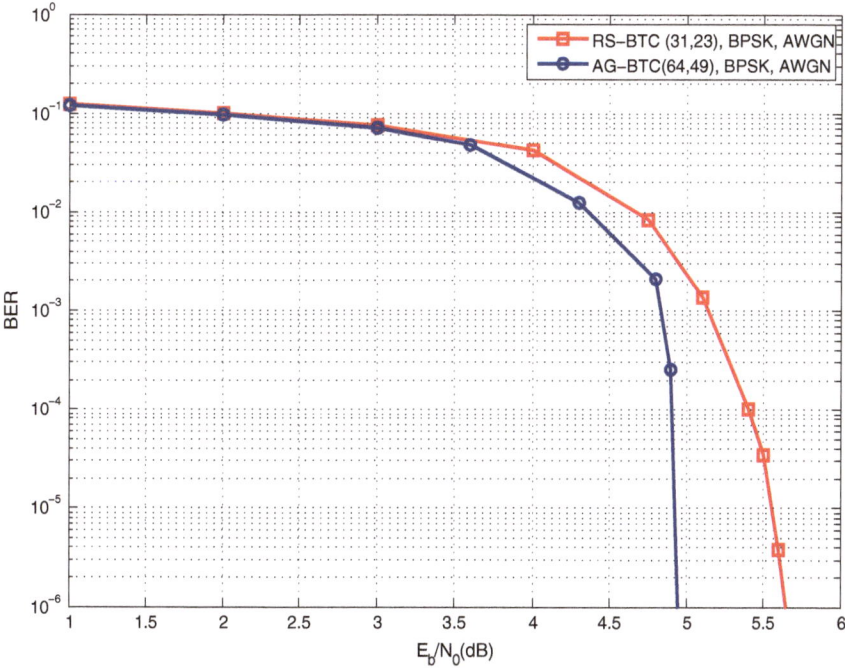

Fig. 4.4 BER of AG-BTC(64,49) versus RS-BTC(31,23) using BPSK over AWGN

gain over various channel models using different modulation schemes. This will allow us to have a comprehensive investigation and gives insights into the impact of various parameters. For example, the selection of the number of the LR bits offers an interesting trade-off between performance and complexity. Thus, the optimum number of LR bits is obtained by finding the maximum number that results in the best BER performance after which the performance improvements are negligible at higher complexity cost. This optimum number obtained from numerical simulation was found to be 4. However, we intentionally did not show the BER performance for each iteration for the sake of keeping the figures as neat as possible in all comparisons that involved BTCs of both AG and RS codes.

Simulations comparing the performances of AG-BTC and RS-BTC codes were carried out. For all modulation schemes and across different code rates and channel models, the superiority of AG-BTC was clearly demonstrated. These results are shown in Figs. 4.4, 4.5 and 4.6 for BPSK modulation over an AWGN channel. The coding gain of AG-BTCs at finite field $GF(2^4)$ for BER of 10^{-6} are 0.7, 0.92 and 1.22 dBs with code rates of 0.59, 0.47 and 0.37, respectively, in comparison to RS-BTC of code rate 0.55 at finite field $GF(2^8)$. Those gains are clearly much higher than those obtained from AG code itself.

It is worth observing that even though the code rate of the AG-BTC at rate 0.59 is higher than the code rate of RS-BTC at 0.55, there is still significant coding gain

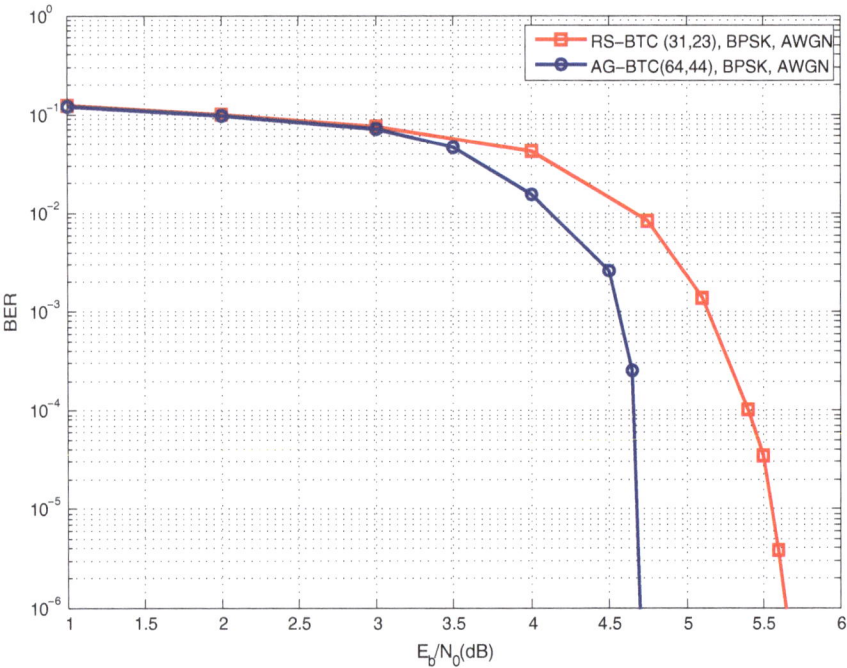

Fig. 4.5 BER of AG-BTC(64,44) versus RS(31,23) using BPSK over AWGN

achieved. This versatility of AG codes stemming from the fact it can be used in concatenation in a BTC process indicates their usability for a wide range of applications where the RS-BTC is preferred. However, such flexibility in getting higher coding gains comes at the cost of slightly higher system complexity due to the use of Chase-Pyndiah's algorithm. From the channel capacity perspective, the AG-BTCs result in 0.3, 0.354 and 0.361 bits per channel use shift from the Shannon capacity at BER 10^{-6} for code rates 0.59, 0.47 and 0.37, respectively, whereas the RS-BTC is 0.365 bits per channel use shift from the Shannon capacity at same BER and code rate of 0.55.

Similarly for QPSK modulation scheme over AWGN channel, coding gains of AG-BTCs at BER of 10^{-6} are 0.7, 1.05 and 1.35 dBs with code rates of 0.59, 0.47 and 0.37, respectively in comparison to RS-BTC of code rate 0.55 all at the same finite field lengths as the BPSK modulation simulations. Those results are shown in Figs. 4.7, 4.8 and 4.9. As expected, the coding gain difference between BPSK and QPSK is minimal.

For the 16QAM modulation scheme over AWGN channels, gains are more significant especially at lower code rates. Coding gains of 1.1, 1.6 and 2.3 dBs are achieved at BER of 10^{-6} with code rates equal to the QPSK code rates. These gains are shown in Figs. 4.10, 4.11 and 4.12.

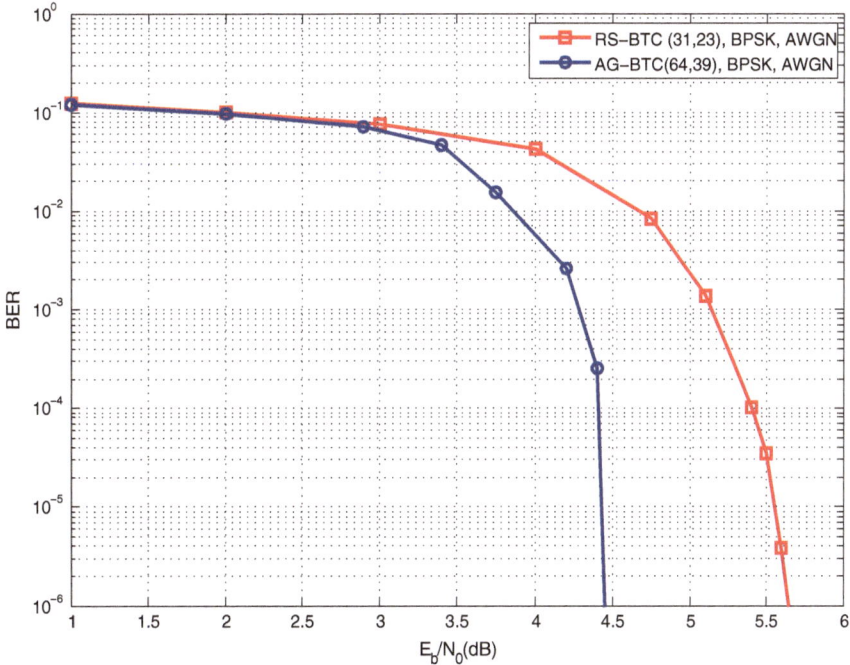

Fig. 4.6 BER of AG-BTC(64,39) versus RS-BTC(31,23) using BPSK over AWGN

Considering the highest modulation scheme, 64QAM allows us to examine the performance gains at higher probability of channel error rate over an AWGN channel. The obtained coding gains are 1.8, 2.45 and 3.3 dBs at BER of 10^{-6} with code rates equal to the 16QAM code rates. These gains are shown in Figs. 4.13, 4.14 and 4.15. We note that the coding gains increase as the modulation index increases. This is of particular importance in next-generation communications systems requiring high throughput and reliability.

An evaluation over Rayleigh fast fading channel was also carried out. A fast fading model is employed in which the coherence time (τ) is far less than the system maximum codeword length. In particular we set the coherence time to 1 bit duration. This represents the worst case scenario and allows us to obtain the lower bound on the coding gain. It also tests the effectiveness of the AG-BTCs over various modulation schemes.

Simulation results compare the BER performance of AG-BTCs with code rates 0.59, 0.47 and 0.37 and RS-BTC with code rates 0.55 for BPSK, QPSK, 16QAM and 64QAM modulation schemes. Gains are clearly much higher than the ones obtained over an AWGN channel. This illustrates that most improvements from the AG-BTCs design is achieved at extreme channel conditions. This is very appealing to next generation wireless systems employing orthogonal frequency division multiple

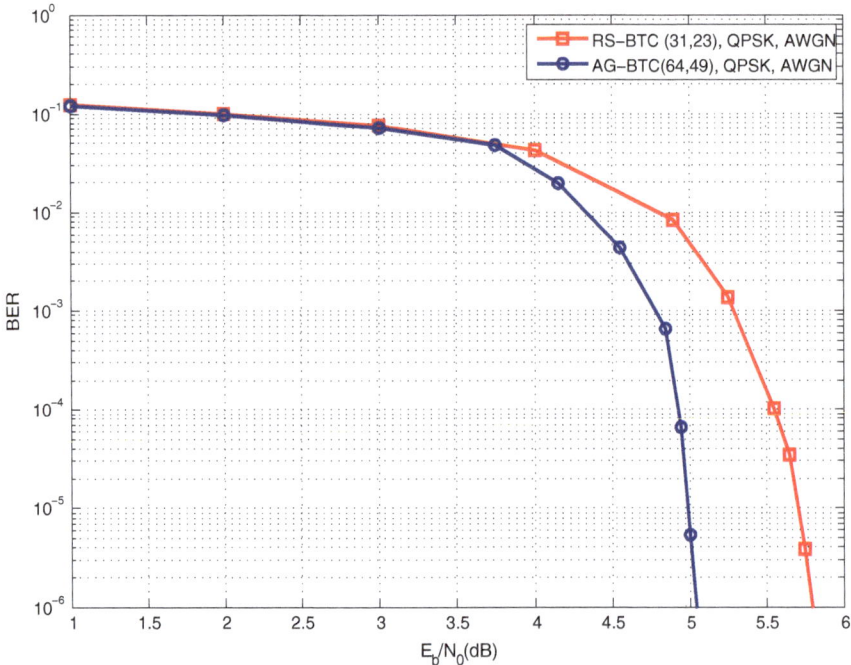

Fig. 4.7 BER of AG-BTC(64,49) versus RS-BTC(31,23) using QPSK over AWGN

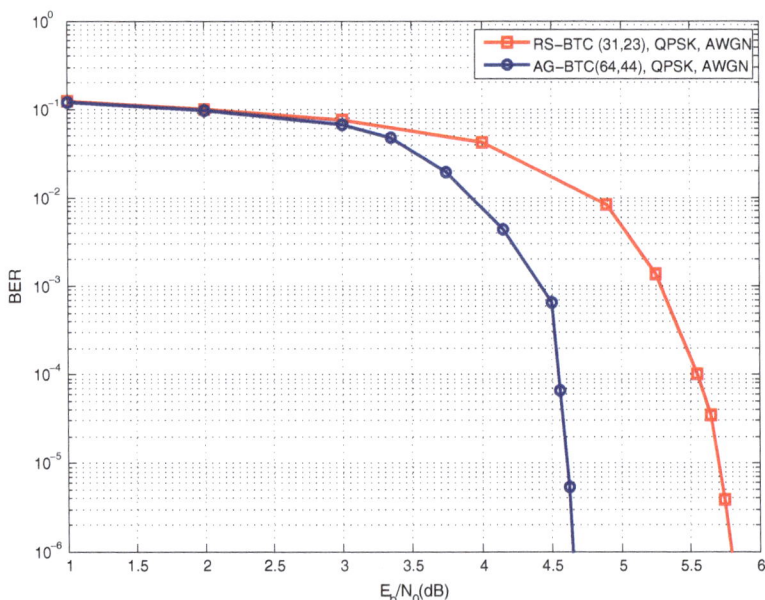

Fig. 4.8 BER of AG-BTC(64,44) versus RS-BTC(31,23) using QPSK over AWGN

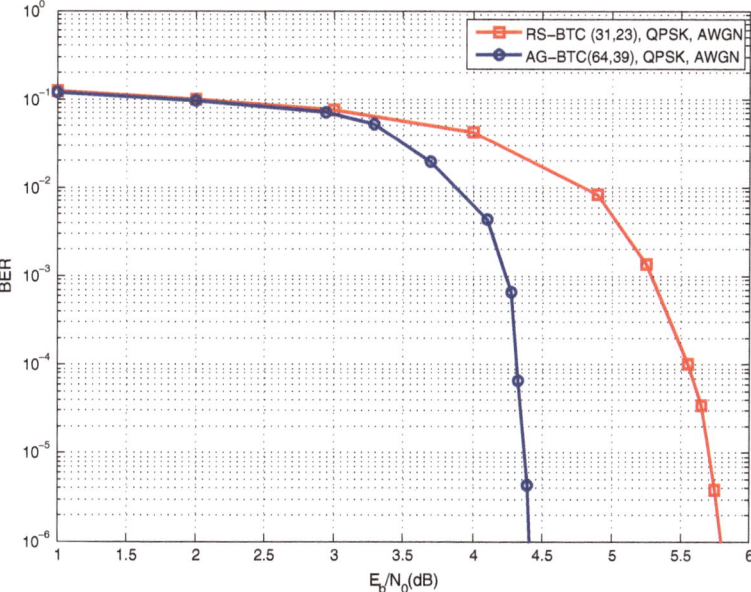

Fig. 4.9 BER of AG-BTC(64,39) versus RS-BTC(31,23) using QPSK over AWGN

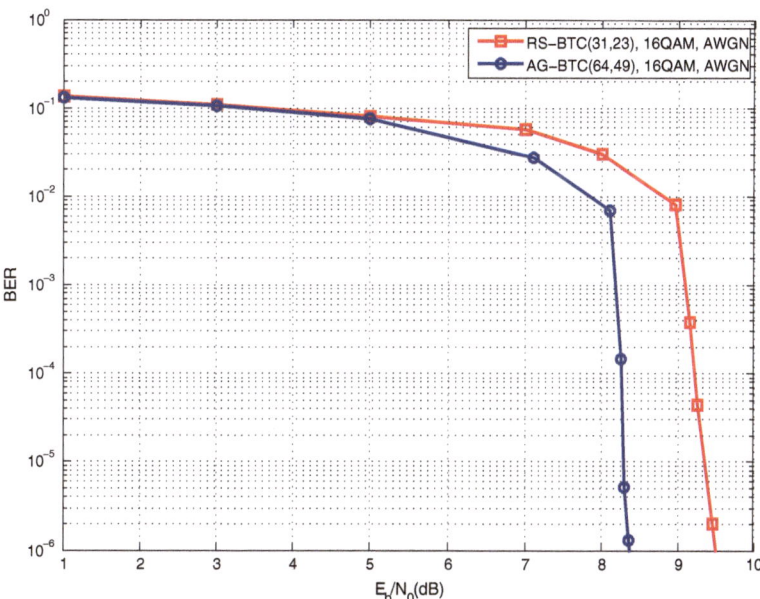

Fig. 4.10 BER of AG-BTC(64,49) versus RS-BTC(31,23) using 16QAM over AWGN

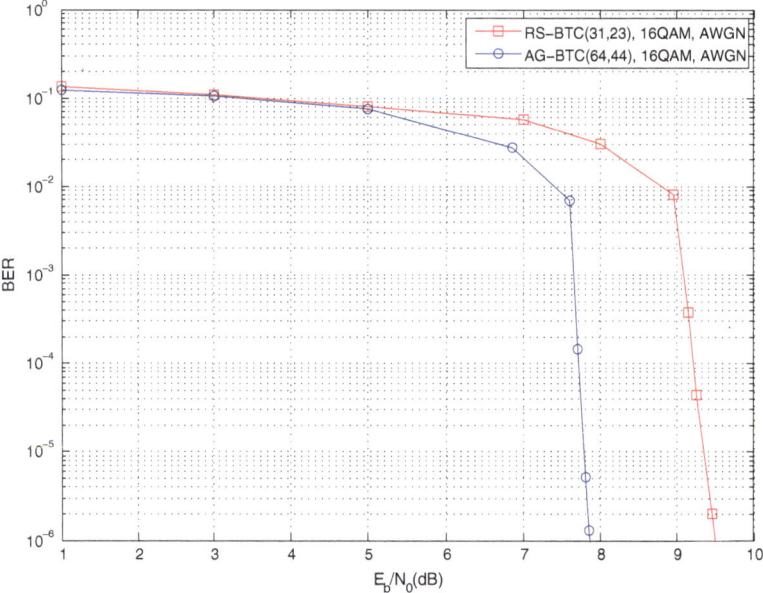

Fig. 4.11 BER of AG-BTC(64,44) versus RS-BTC(31,23) using 16QAM over AWGN

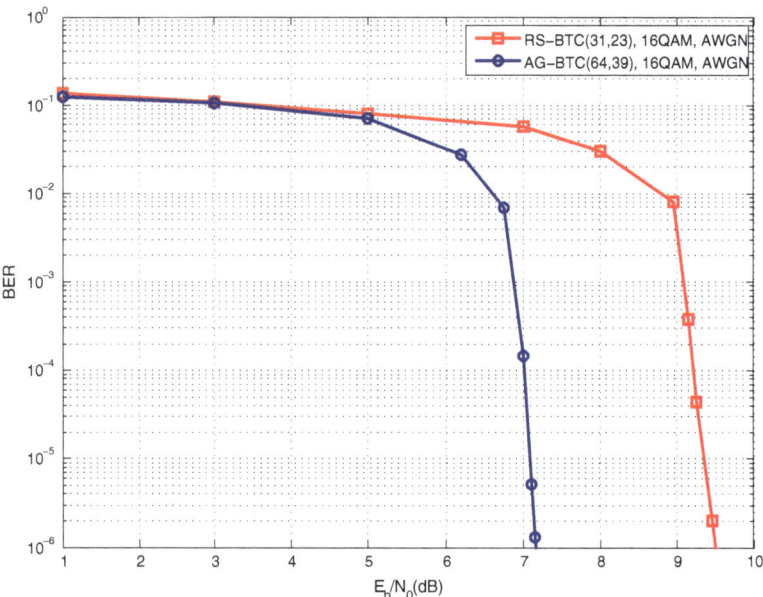

Fig. 4.12 BER of AG-BTC(64,39) versus RS-BTC(31,23) using 16QAM over AWGN

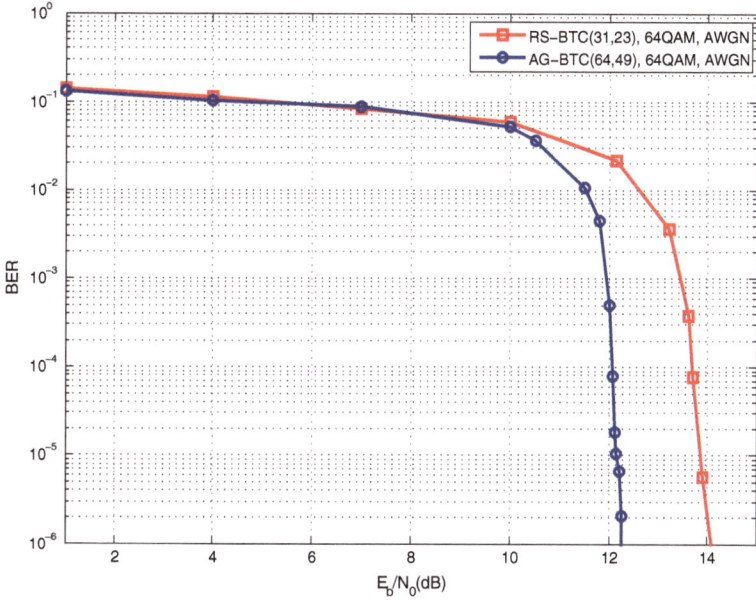

Fig. 4.13   BER of AG-BTC(64,49) versus RS-BTC(31,23) using 64QAM over AWGN

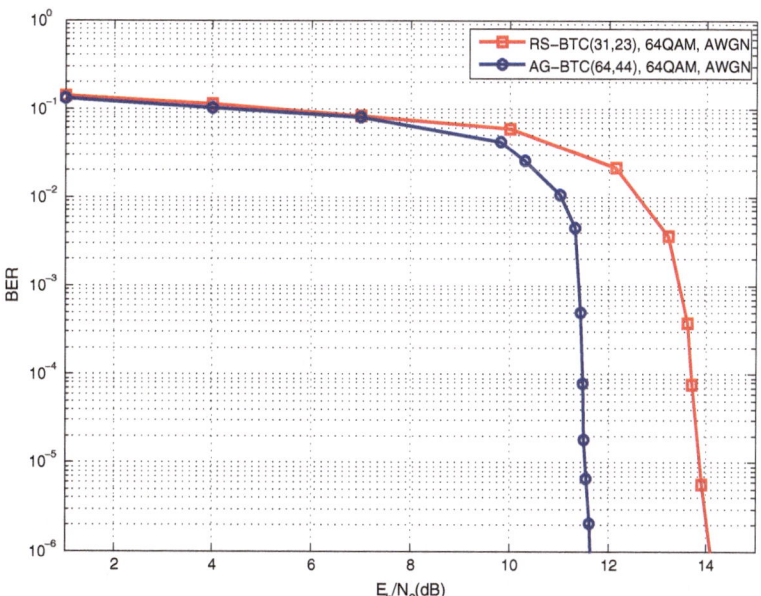

Fig. 4.14   BER of AG-BTC(64,44) versus RS-BTC(31,23) using 64QAM over AWGN

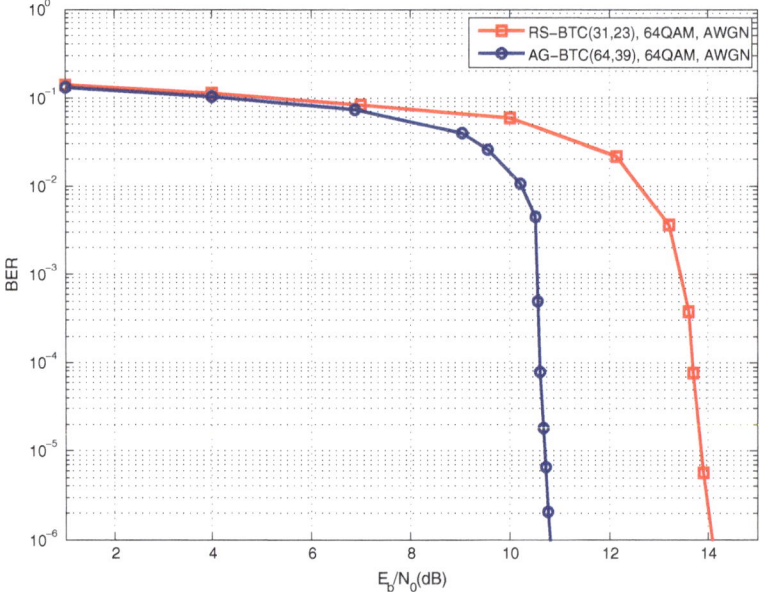

Fig. 4.15   BER of AG-BTC(64,39) versus RS-BTC(31,23) using 64QAM over AWGN

access (OFDMA) where fading is a major problem in the way of achieving the expected high throughputs. For more detailed results, the reader is referred to Alzubi's PhD dissertation [8].

4.4 Summary

In this chapter, we first wrote code for computer simulations to evaluate the BER performance of AG codes and compare with the performance of RS codes. Simulation results confirmed the correctness of the developed software platform by matching exactly published results in the literature for the case of BPSK modulation over both AWGN and Rayleigh fast fading channel conditions.

The design, construction and implementation of AG-BTCs are presented. For BPSK modulation over AWGN channel model, results show coding gains of 0.7, 0.92 and 1.22 dBs for the AG codes of code rates 0.59, 0.47 and 0.37 respectively over the RS code of code rate 0.55. A slight increase in coding gain is observed for the case of QBPSK modulation. For 16QAM, coding gains of 1.1, 1.6 and 2.3 dBs for the AG codes of code rates 0.59, 0.47 and 0.37 respectively over the RS code of code rate 0.55. Those gains are 0.4, 0.32 and 0.92 dBs more than the gains obtained using BPSK modulation for the same code rates and channel model. Similarly for 64QAM over an AWGN channel, the achieved gains are 1.1, 1.53 and 2.08 dBs more than the gains obtained using BPSK modulation.

Those results are encouraging and show the applicability of AG codes as a code component of BTCs. This combination is useful and could work well in applications such as video transmission. The trend of increased coding gains with the modulation index increase is clearly noticeable.

Simulation results in this chapter highlight the benefits of using AG codes as a code component of BTCs over RS codes of same structure using different modulation schemes and over AWGN channel. However, this comes at the cost of increased overall system complexity owing to using Chase-Pyndiah's decoding along with AG codes in the case of BTC. This problem is addressed in the next chapter.

References

1. Gallager RG (2008) Principles of digital communication. Cambridge University Press, Cambridge. http://books.google.co.uk/books?id=5W0aYFU02igC
2. Rappaport T (2001) Wireless communications: principles and practice, 2nd edn. Prentice Hall PTR, Upper Saddle River, NJ, USA
3. Aitsab O, Pyndiah R (1996) Performance of Reed-Solomon block turbo code. In: IEEE GLOBE-COM'96, vol 1, pp 121–125. doi:10.1109/GLOCOM.1996.594345
4. Awad A (2007) Detection and coding techniques for fourth generation air-interfaces based on multicarrier modulation. Ph.D. thesis, The University of Leeds, Leeds, UK
5. Pyndiah R, Picart A, Glavieux A (1995) Performance of block turbo coded 16-qam and 64-qam modulations. In: IEEE GLOBECOM'95, vol 2, pp 1039–1043. doi:10.1109/GLOCOM.1995.502563
6. Mahran A, Benaissa M (2003) Adaptive chase algorithm for block turbo codes. Electron Lett 39(7):617–619. doi:10.1049/el:20030421
7. Pyndiah RM (1998) Near-optimum decoding of product codes: block turbo codes. IEEE Trans Commun 46(8):1003–1010. doi:10.1109/26.705396
8. Alzubi J (2012) Forward error correction coding and iterative decoding using algebraic geometry codes. Ph.D. thesis, Swansea University, Swansea, Wales

Chapter 5
Irregular Decoding of Algebraic-Geometric Block Turbo Codes

In the previous chapter, the design, construction, and implementation of quantam AG-BTCs were proposed and investigated in depth. The designed system suffered from high complexity in the decoding side due to the use of Chase-Pyndiah's decoding algorithm. In this algorithm, the decoding process complexity is exponentially related to the number of LR bits chosen.

To overcome this drawback of the designed codec, a new design and construction method of IBTCs is proposed in this chapter. The new design is inspired by the idea of unequal protection of information symbols which is central to IBTCs.

The chapter starts with presenting the design and construction method of the AG-IBTC. A design for the AG-IBTC decoder is proposed with detailed explanation of the decoding process. Simulations results for BER performance of the new AG-IBTC are presented and compared with the BER performance results of equivalent AG-BTCs.

Finally this chapter will be concluded with observations about the gain obtained by implementing the proposed design and the complexity reduction achieved.

5.1 Irregular AG Block Turbo Code Encoder

The conventional encoding method of BTCs or turbo product codes (TPCs) is summarised here. The information bits are arranged in a block format, and then passed into a systematic block encoder which is nothing more than multiplication of the information bits (block) by a systematic generator matrix constructed according to a certain set of rules depending on the type of the code being used. The output is then interleaved using a block interleaver which converts the rows into columns and vice versa. The output from the block interleaver is passed through another systematic block encoder of the same type as the first one [1]. Figure 5.1 shows the described encoding method.

J. A. Alzubi et al., *Forward Error Correction Based On Algebraic-Geometric Theory*, SpringerBriefs in Electrical and Computer Engineering, DOI: 10.1007/978-3-319-08293-6_5, © The Author(s) 2014

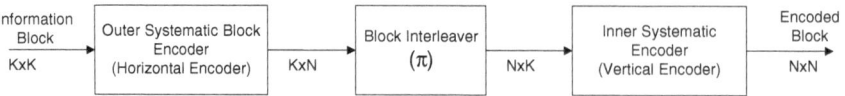

Fig. 5.1 Conventional block encoding method

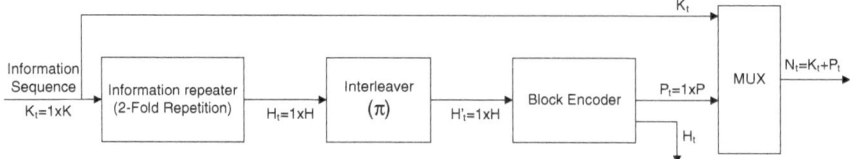

Fig. 5.2 Equivalent irregular encoding structure for TPCs

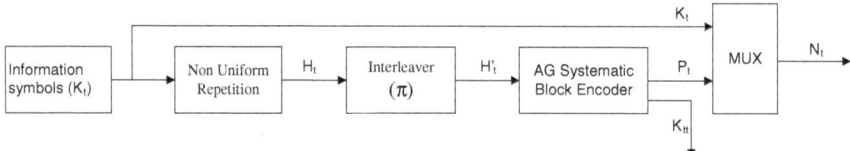

Fig. 5.3 Encoding structure for irregular BTC

However, there is an equivalent structure of the TPCs which is composed of only one encoding component in contrast with the conventional encoding method. In this equivalent method, the information bits are repeated (i.e., twice) in the case of TPCs of even information bits. The reason is because each information bit will have two extrinsic information values in the decoding process, one from the outer decoding component and the other from the inner decoding component. Similarly in $2°$ IBTC, every information bit in the codeword will have two extrinsic information values in the decoding process [2]. This equivalent structure is illustrated in Fig. 5.2.

The core idea of the equivalent structure of TPCs mentioned above is applied to design and construct the AG-IBTC. Assume degree d is the number of times that a fraction of information (non-binary symbols) is repeated with a restriction that $d \geq 2$. The higher the value of d, the stronger the protection on the symbols as the a posteriori value of those symbols will be derived from d number of extrinsic information symbols.

Figure 5.3 illustrates a block diagram of the IBTC encoder. The information symbols to be encoded K_t will be passed into a non-uniform repetition unit which will splits the information symbols into j groups, where j should not exceed 3 for a good code performance. Each group is repeated d_i times where $d_i = 2, 3, \ldots, T$, and T is the maximum number of repetitions. A fractions of the total information symbols f_j is the number of symbols in a group j, where

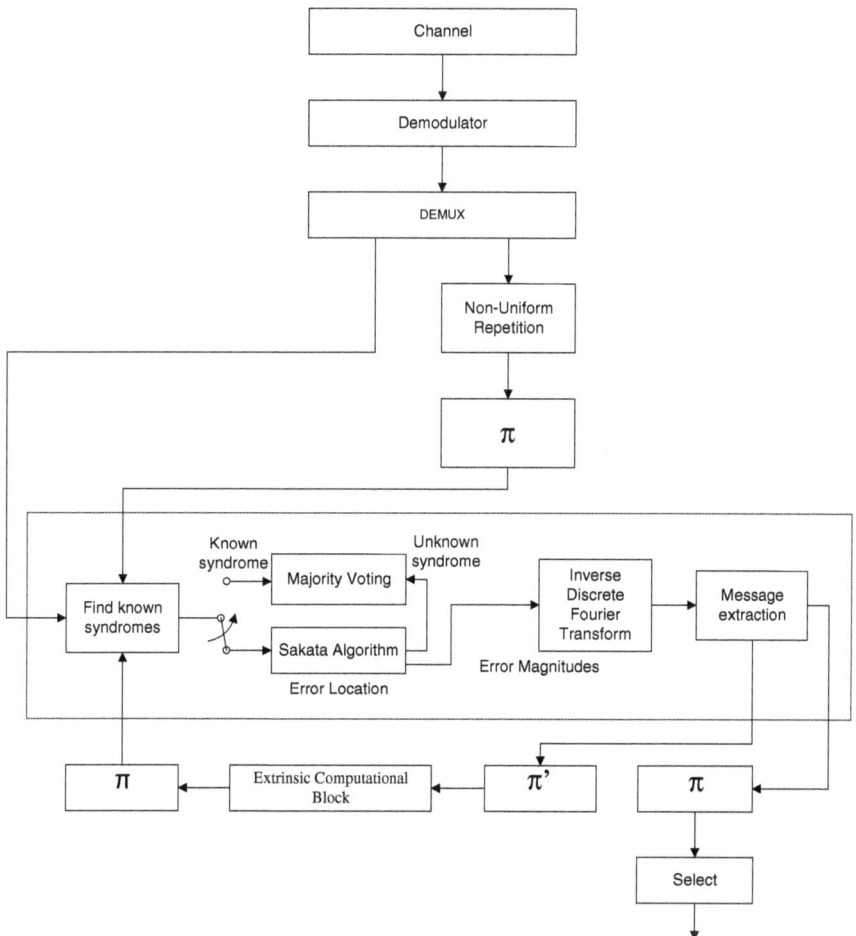

Fig. 5.4 Decoder structure for irregular BTC

$$\sum_{j=1}^{3} f_j = K_t \tag{5.1}$$

Hence, f_i is repeated d_i times by the non-uniform repetition unit in the encoder. The information after being processed in the non-uniform repetition unit will be:

$$H_t = \sum_{i=2}^{T} \sum_{j=1}^{3} d_i \cdot f_j \tag{5.2}$$

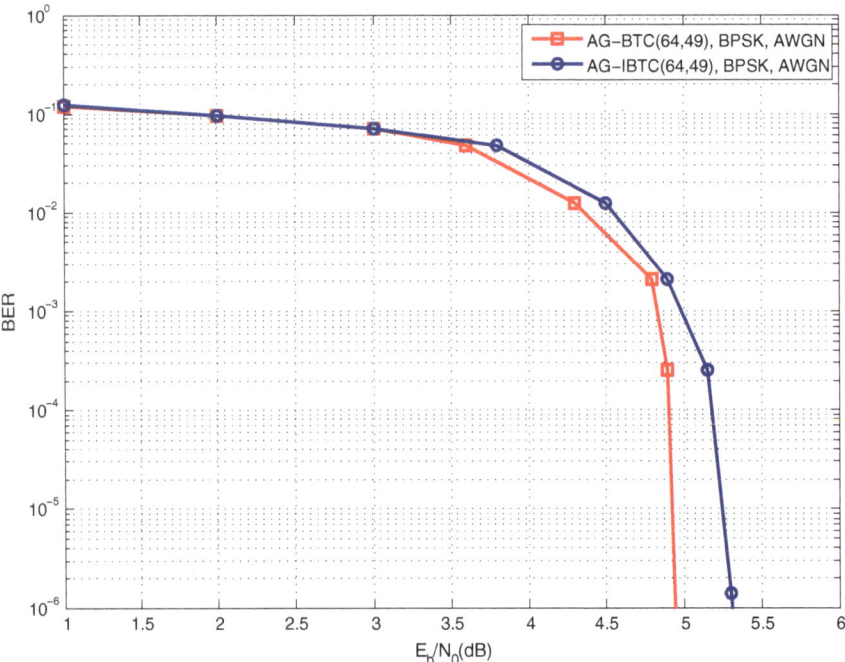

Fig. 5.5 BER of AG-IBTC(64, 49) versus AG-BTC(64, 49) using BPSK over AWGN

The length of the code (codeword size) controls and limits the options of choosing an appropriate symbol degree d_i and a corresponding fraction f_j. Further explanation will be presented later in this section. Thus far there is no known algorithm that computes an optimal combination of these values. However, the symbol degree profile is preferably to contain a fraction f_j of the information symbols repeated one time (degree 2). Generally, the symbol degree two preferred to have a fraction f_j between 75 and 95 % of the original information symbols, while higher degrees share the remaining fraction depending on the modulation scheme used [2].

Using the above criteria in designing the AG-IBTC, one of the codes used here is AG-IBTC(64, 49) with the symbol degree combination of 85 % fraction of the information block repeated once (degree 2), 10 % fraction is repeated twice (degree 3), and 5 % fraction is repeated eight times (degree 9).

The output of the non-uniform repeating unit in Fig. 5.3 is then interleaved randomly using a random interleaver and then passed into an AG systematic block encoder. The parity bits P_t can be easily extracted from the output of the AG systematic encoder to be appended to the original information K_t before the non-uniform repetition unit to be transmitted together in block format N_t. The code rate of such IBTC is:

$$R = \frac{K_t}{K_t + P_t} \tag{5.3}$$

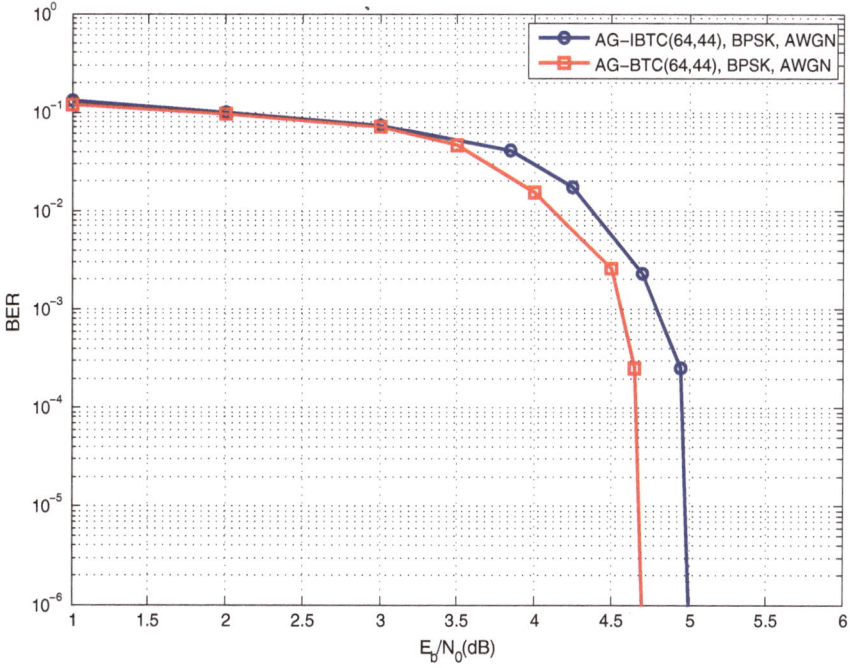

Fig. 5.6 BER of AG-IBTC(64, 44) versus AG-BTC(64, 44) using BPSK over AWGN

where R represents the code rate of the IBTC, K_t is the length of original information, and P_t is the length of parity.

As previously mentioned in this section, the degrees d_i play an important role in the construction of the IBTC but are limited and controlled by the block size of the corresponding regular BTC for a fair comparison in terms of the BER performance. The following example will illustrate this idea.

Considering one of the AG non-binary BTC codes used here, a systematic AG non-binary BTC (n, k, d) where n and k are the lengths of the codeword and the information in non-binary symbols, respectively, and d is the minimum Hamming distance. Thus in designing a corresponding AG non-binary I-BTC, the size of the information to be encoded and transmitted, K_t, which will be grouped in terms to be repeated must be equal to the size of the information k in the regular AG non-binary BTC [3].

For instance, the AG non-binary I-BTC derived from a $(64, 49, 10)$ systematic AG non-binary regular BTC could have a block size of information K_t of 49×20 non-binary symbols, for each row k_t in K_t using a 2° of repetition for 85 % fraction f_j of the information, 3° for 10 % fraction f_j, and degree 9 for 5 % fraction f_j of the information, where $j = 1, 2, 3$. These combinations will produce a repeated information block H_t of size 49×49 where each row is called h_t. Although there exist a few other combinations, which will ensure that H_t retains the original dimension

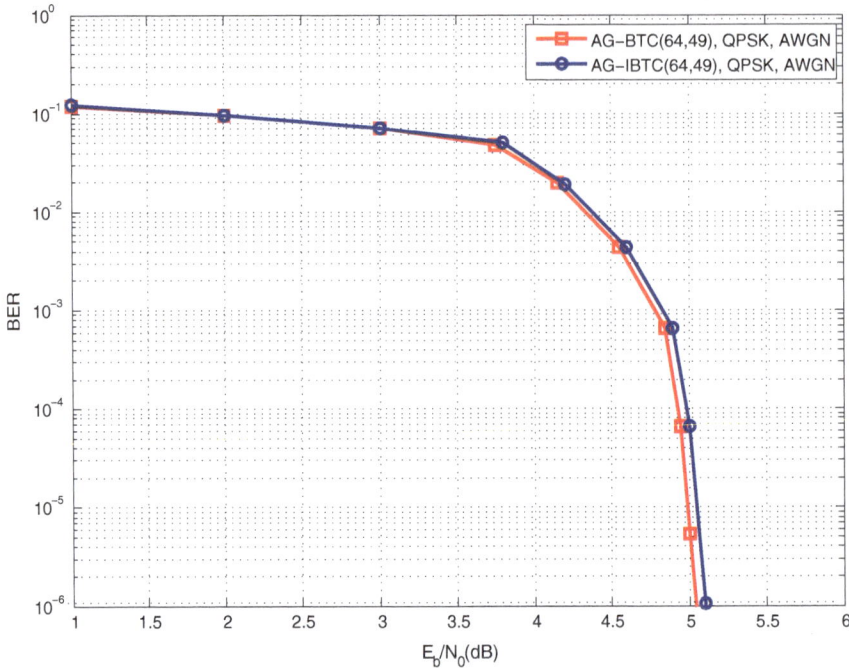

Fig. 5.7 BER of AG-IBTC(64, 49) versus AG-BTC(64, 49) using QPSK over AWGN

of the regular BTC information block before encoding, but the just described combination showed the optimum BER performance with comparison to $(64, 49, 10)$ systematic AG non-binary regular BTC.

A random interleaver is used to interleave the whole array H_t, and then each row from the interleaved version of H_t which is called H_t' will be read out individually as a vector h_t' and then encoded using the AG systematic encoder separately. Parity non-binary symbols are then extracted from the encoded vector while the information part is discarded. The parity vector P_t is then attached to the original information symbols vector k_t to form the encoded message n_t. A collection of encoded messages compose a block of encoded information symbols N_t in order to be modulated and transmitted via the channel. The rate of the produced code $R = 0.57$ is almost similar to the rate of the equivalent regular BTC $(64, 49, 10)$ which is $R = 0.585$.

5.2 Irregular AG Non-binary Block Turbo Code Decoder

The received encoded block N_r will be demodulated using a proper demodulator, and then passed through a demultiplexer in order to separate the parity P_r symbols in each encoded message n_r from the information symbols k_r. The information part

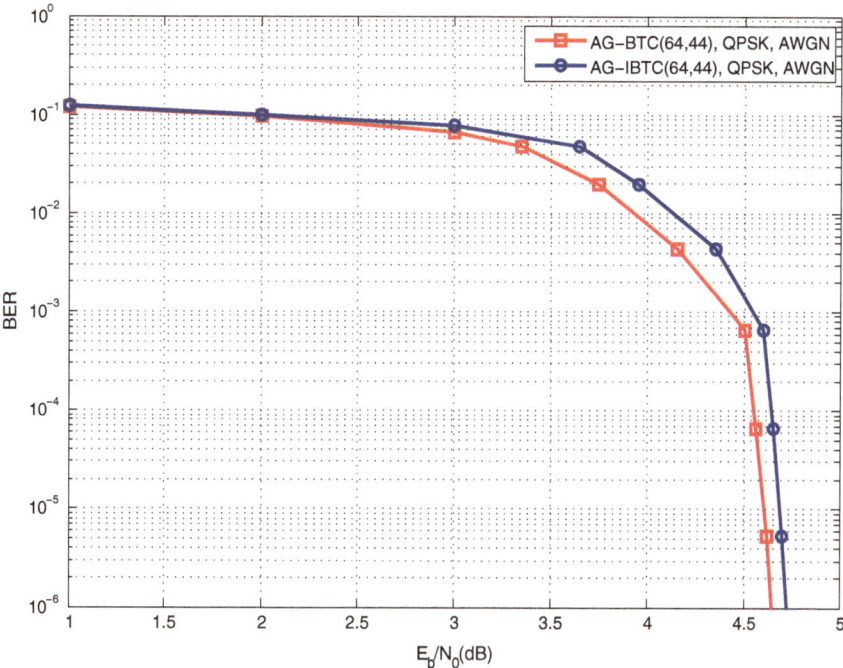

Fig. 5.8 BER of AG-IBTC(64, 44) versus AG-BTC(64, 44) using QPSK over AWGN

K_r is then repeated in the same manner as at the transmitter side using the same non-uniform repetition unit to produce a repeated information block H_r in order to interleave it randomly using the same random interleaver being used at the transmitter side [2, 4]. Each parity vector P_r is then attached to its corresponding interleaved vector of repeated information h'_r to form a vector n_r. An initial a priori value a_r of equal probability (i.e., zero log-likelihood) will be added to the vector n_r before entering the AG decoder.

Extrinsic information e_r of same size as vector h_r is collected and computed from the output of the AG decoder. The block of extrinsic information E_r is deinterleaved using a random deinterleaver before being passed into the extrinsic computational block [4]. A new extrinsic information value is computed for every information symbol of degree d_i at every iteration, and this new extrinsic information value is the product of the other $d_i - 1$ extrinsic information values or is the sum of those values when using log-likelihood values [3].

The new a priori values block A_r is set by randomly interleaving the output of the extrinsic computational block which is of same size as the extrinsic information block E_r. The a priori vectors a_i are then read out individually in order to be added to n_r vectors for the next decoding iteration.

After final decoding iteration of each codeword, the decoded codeword is stored to form a block of decoded codewords. The block of decoded codewords is then

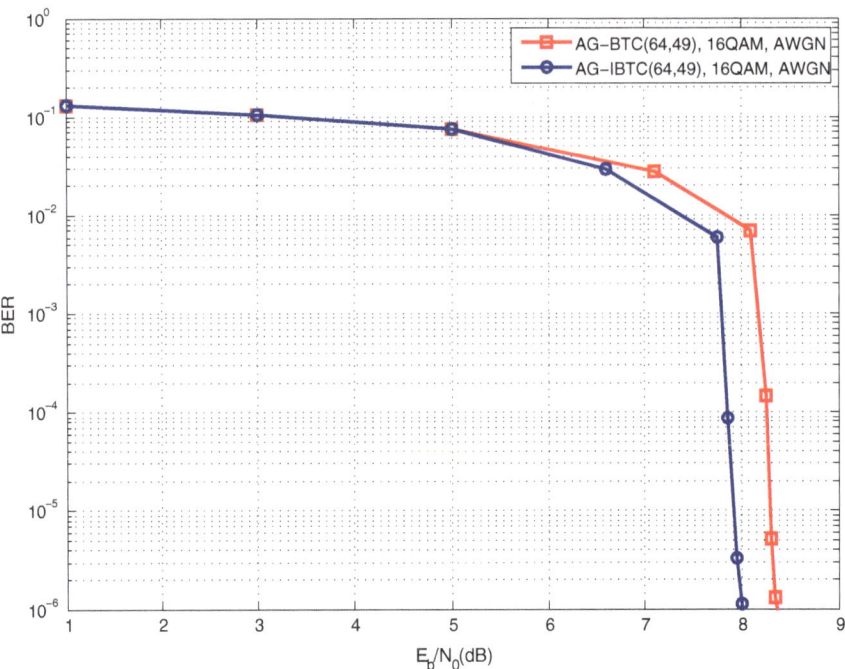

Fig. 5.9 BER of AG-IBTC(64, 49) versus AG-BTC(64, 49) using 16QAM over AWGN

deinterleaved using a random deinterleaver in order to remove the parity part from each codeword. The aim of extracting the information part only from the decoded codeword is to retain the original generated information vector k_t format for comparison purposes [2].

The whole decoding process is illustrated in Fig. 5.4. It should be noted that the terms random interleaver and random deinterleaver implies that the randomness in the interleaver and deinterleaver is preserved for every data block [4]. In other words the random interleaver and random deinterleaver patterns used for one data block are totally different from those who are used for any other data block.

5.3 BER Performance of AG Irregular Block Turbo Codes Versus AG Block Turbo Codes

In the previous chapter, the AG-BTCs have shown better BER performance compared to RS-BTCs. However, their complexity is relatively high due to the use of Chase-Pyndiah's algorithm for extracting the soft output needed for the iterative process in the AG-BTCs. The design of AG-IBTCs proposed in this chapter could help greatly

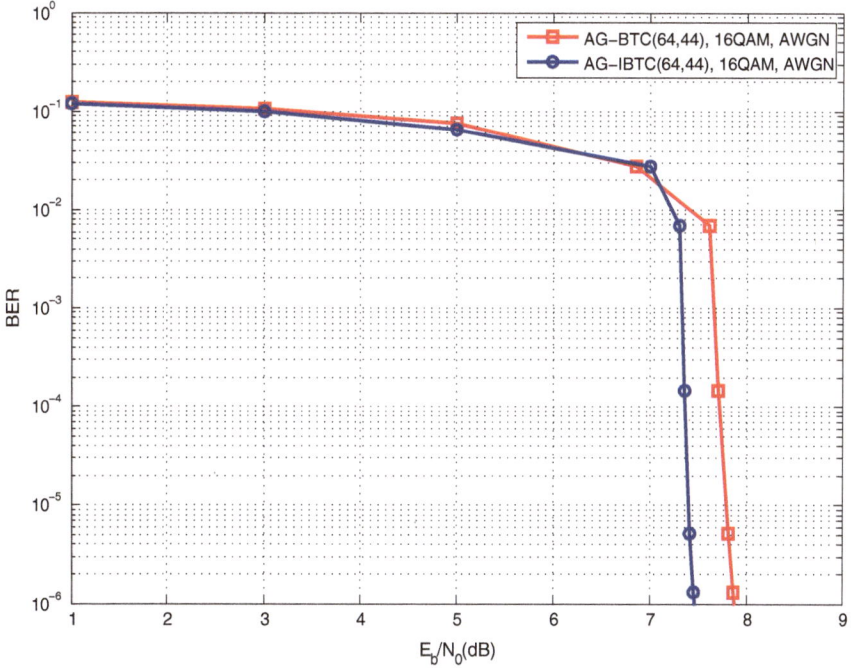

Fig. 5.10 BER of AG-IBTC(64, 44) versus AG-BTC(64, 44) using 16QAM over AWGN

to reduce the system complexity due to using one encoder and one decoder instead of two of each component in the AG-BTC case.

The overall AG-IBTC system complexity is still less than the AG-BTC despite requiring more iterations for the same BER performance. Our aim is to explore this complexity-performance trade-off and highlight the conditions under which AG-IBTC outperforms AG-BTC while keeping the complexity at minimum. The results for different modulation schemes and channel models are presented in this chapter (the BER performance for each iteration were intentionally not shown for the sake of keeping the figures as neat as possible).

For the sake of fair comparison between the AG-IBTC and AG-BTC, similar data block sizes and almost the same code rates are chosen over the same finite field $GF(2^4)$. It is important to mention that the optimal combination of the AG-IBTC has been obtained from simulations and selected to be used in the comparison with AG-BTC. Using BPSK modulation over an AWGN channel as shown in Figs. 5.5 and 5.6, the coding gains in BER performance of AG-IBTC codes at BER of 10^{-6} are -0.35 and -0.27 dBs with code rates of 0.57 and 0.5 respectively in comparison to AG-BTC codes of code rates 0.585 and 0.47. The losses from using the AG-IBTCs design are negligible given the significant reduction in the system complexity.

Figures 5.7 and 5.8 show the QPSK results over an AWGN channel. The coding gains in BER performance of AG-IBTC codes at BER of 10^{-6} are -0.1 and

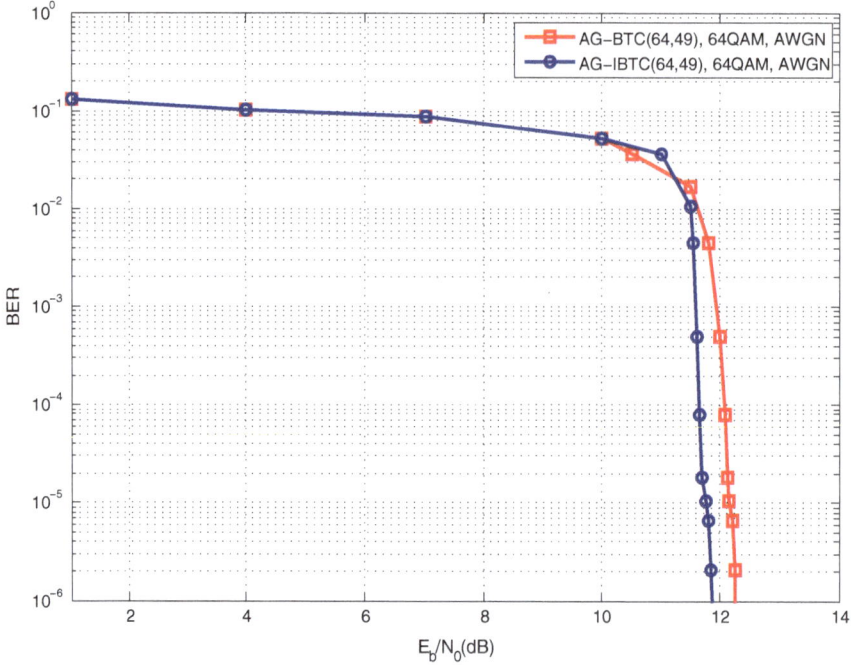

Fig. 5.11 BER of AG-IBTC(64, 49) versus AG-BTC(64, 49) using 64QAM over AWGN

−0.08 dBs with code rates of 0.57 and 0.5 respectively in comparison to AG-BTCs of code rates 0.585 and 0.47. Although the losses are again negligible, they are decreasing at faster rate as the modulation index increases to QPSK.

Figures 5.9 and 5.10 show the 16QAM results over an AWGN channel. The coding gains in BER performance of AG-IBTC codes at BER of 10^{-6} are 0.35 and 0.4 dBs with code rates of 0.57 and 0.5 respectively in comparison to AG-BTCs of code rates 0.585 and 0.47. It can be seen that the gains are positive. Not only is system complexity reduced but BER performance gains are also achieved. This is consistent with the gain improvement trend for AG-BTCs codes as the modulation index increases that was highlighted earlier.

Figures 5.11 and 5.12 show the 64QAM results over an AWGN channel. The coding gains in BER performance of AG-IBTC codes at BER of 10^{-6} are 0.4 and 0.55 dBs with code rates of 0.57 and 0.5 respectively in comparison to AG-BTCs of code rates 0.585 and 0.47. This is the point at which the highest BER performance gain and large reduction in system complexity are achieved. Further, we can emphasize the adaptability of AG codes in various coding design with BTC and IBTCs.

The same code rates, finite field, and data block sizes as in the AWGN channel model were re-used for the Rayleigh fast fading channel. Losses from using the AG-IBTCs design seems considerable at first glance. However, it can be observed

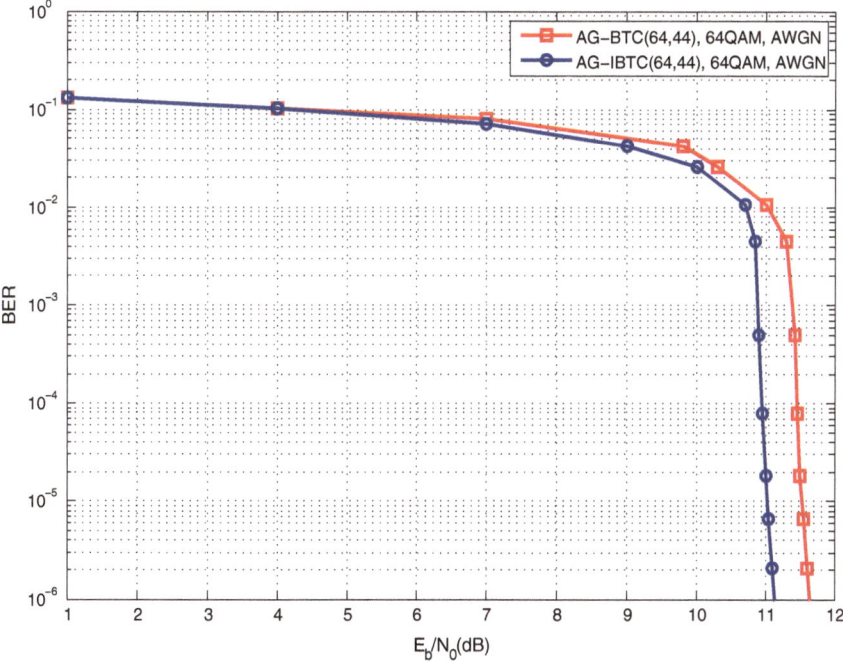

Fig. 5.12 BER of AG-IBTC(64, 44) versus AG-BTC(64, 44) using 64QAM over AWGN

that these losses are applicable to the high *EB/No* region and come as the result of severe fading channel conditions. Moreover, AG-IBTC has the benefit of substantial reduction in system complexity which is highly desirable in severe fading conditions. For more detailed results, the reader is referred to Alzubi's Ph.D. dissertation [5].

5.4 Summary

In order to overcome the high system complexity of AG-BTCs, a solution based on the IBTC is proposed. This approach can substantially reduce system complexity while maintaining BER performance. Simulations were carried out in Matlab to measure the BER performance of AG-IBTCs and compare to their equivalent AG-BTCs over AWGN and Rayleigh fast fading channel models. The comparison is performed on similar data block length, code rates and over the same finite field.

For BPSK modulation, AG-IBTC results in 0.35 and 0.27 dBs coding loss at BER of 10^{-6} for code rates 0.57 and 0.5 respectively over AWGN channel. For QPSK again the coding loss is a bit lower than the BPSK case. For both cases and despite the coding loss, a significant system complexity reduction is obtained which is clearly shown in the design and construction of the AG-IBT codec. For 16QAM and 64QAM, the coding gains become positive, i.e., 0.35 and 0.6 dBs and 0.4 and 0.8 dBs for code rates 0.57 and 0.5, respectively, at BER of 10^{-6}. Such transition from negative to positive confirms the fact that the AG codes in general and specifically AG-IBTCs gain are better when the modulation index increases. Also it gives a solution to the complexity issue of AG-BTC.

References

1. Soleymani MR, Gao Y, Vilaipornsawai U (20012) Turbo coding for satellite and wireless communications. No. 1 in Kluwer international series in engineering and computer science. Kluwer Academic Publishers. http://books.google.co.uk/books?id=r6yfH6Z69NsC
2. Sholiyi A (2011) Irregular block turbo codes for communication systems. Ph.D. thesis, Swansea University, Swansea, UK
3. Richardson TJ, Shokrollahi MA, Urbanke RL (2001) Design of capacity-approaching irregular low-density parity-check codes. IEEE Trans Inf Theor 47(2):619–637. doi:10.1109/18.910578
4. Frey B, Mackay D (1990) Irregular turbo codes. In: 37th Allerton Conference on Communication, Control and Computing. Allerton House, Illinois
5. Alzubi J (2012) Forward error correction coding and iterative decoding using algebraic geometry codes. Ph.D. thesis, Swansea University, Swansea, Wales

Chapter 6
Conclusions

In this book, BER performance results for AG codes published in the literature have been verified through simulations. Those results allow us to draw solid conclusions about the performance, parameterization, and characteristics of AG codes. For both AWGN and Rayleigh fast fading channels, results have shown that the AG codes outperform RS codes of the same data block length and rate but over different finite fields due to the nature and construction of each kind of codes. This is a more appropriate and fair comparison than the ones used in the literature.

AG codes seem to offer even higher coding gains in Rayleigh fast fading channels than AWGN channels. In addition, the coding gain is directly proportional to the modulation index which suggests that they offer more resilience to adverse channel conditions currently impeding throughput in wireless networks. Considering the ability to achieve higher coding gains using higher modulation indexes, AG codes seem to be a good candidate technology for next-generation wireless systems.

The approach of using AG codes as code components in design of BTC has shown several benefits and challenges. One challenge was to extract soft outputs from the hard decision outputs of the AG codes as required by the BTC design. This has been addressed via introducing the Chase-Pyndiah method for extracting such output. This resulted in additional system complexity.

The benefits include higher coding gain which was measured by BER performance at different AG-BTC code rates in comparison to an equivalent RS-BTC. In contrast to the literature, our comparison is performed on the basis of same data block length and code rates but different finite fields as mentioned earlier. We believe this is more accurate as the number of simulated bits and code rates matter more than the finite field size. This also helps to keep the effect of the chosen number of LR bits.

Using Matlab simulations, we were able to compare the BER performance using different modulation schemes over AWGN channel. AG-BTCs have outperformed RS-BTCs in all simulated scenarios. The coding gains achieved increase as the modulation index increases. The results show an attractive adaptability of AG codes to change in code design.

J. A. Alzubi et al., *Forward Error Correction Based On Algebraic-Geometric Theory*,
SpringerBriefs in Electrical and Computer Engineering,
DOI: 10.1007/978-3-319-08293-6_6, © The Author(s) 2014

In order to alleviate the problem of high system complexity arising from AG-BTC, we propose the design and construction of AG-IBTC (for the first time to the best of our knowledge). Simulation results comparing AG-IBTCs with equivalent AG-BTCs were obtained and presented here. For the first time, simulations have been carried out comparing regular and irregular BTC based on AG codes using different modulations over AWGN and Rayleigh fast fading channels. Simulation results highlight the trade-off between the BER performance and overall system complexity. It has been shown that in most cases, coding gain is achieved while reducing the system complexity. For a few cases, a negligible coding loss was observed while enjoying a significant system complexity reduction.

We believe the inclusion of AG-IBTCs will offer great flexibility in codec design that is particularly applicable to high-throughput wireless networks that can adjust the trade-off between system complexity and performance. For example, the newly developed AG-IBTC can be exploited in regular BTC applications such as error correction in optical and magnetic storage devices and next generation storage devices such as Blu-rays discs and HD-DVDs, albeit at a reduced complexity.

6.1 Open Research Issues

It has been shown that the BER performance of AG codes is significantly improved compared to RS codes. This is also the case when using AG-BTC. However, the decoding complexity of AG codes is still higher than the complexity of RS codes. The complexity increases when AG codes are used in BTC design. A major component of this complexity is due to the use of Chase-Pyndiah's algorithm to extract soft output from the hard decision outputs of AG codes. This is currently performed on the bit level which requires a large number of computations. Using a symbol-level Chase algorithm will help reduce the complexity of the overall system.

AG-IBTCs have been shown to offer much reduced system complexity while maintaining the BER performance gains of AG-BTCs. Currently, there are no available algorithms to compute the optimal combinations of symbol degree and corresponding fractions to generate IBTCs. Our approach is to find these combinations from computer simulations. A possible goal for future research is a reliable algorithm to compute these combinations.

AG codes have shown significant coding gains improvements as a single code, and as code components of BTCs and IBTCs. Those gains are found to be even higher in severe fading channel conditions while being scalable with the increase in the modulation index. Currently, there are no attempts to include AG codes in wireless communication standards such as OFDMA-based air interface networks (HSPA and LTE) and IEEE WLANs standards such as 802.11g and 802.11n. The reason for this could be due to their high codec complexity in the past. In this book, several techniques for reducing the system complexity are presented and hence further research could be carried out into extending current results to OFDMA and IEEE standards based wireless systems.